Cultural Life at the Abyss

Ideology dominates social research, encouraged by rejections of nature and the past, and often ignores the direct experience of actual people. This archaeological study takes a different approach, grounding concepts of culture, landscape and art in ecological relations that embrace all of life.

An ecological approach considers that life exists in the interactions of people with the environment surrounding them. This theoretical grounding therefore supports research at a local scale and validates the analysis of individual effort. The case studies explore individual perception, action and expression in a startlingly diverse set of objects and features from the past: natural and constructed monuments, ancient and recent rock paintings, petroglyphs, fresco paintings and impressionist landscape art. While traditional cultural approaches render ordinary people as proxies, these individuals, as members of families and communities, do the actual work of society, using their senses, bodies and minds. The analysis here therefore turns away from traditional speculations about the meanings of cultural things to look for evidence of the personal choices of travelers, inhabitants, pilgrims and artists as they acted, and attempt to gain insights from these decisions about the past as lived.

The book will be of interest to scholars, researchers and advanced students in culture and society who may be restless in theatres of discourse dominated by self-affirming narratives, who wish to consider the fields of possibility in an environmental perspective that integrates culture with nature and humans with other beings in a singular, physical world.

B. L. Molyneaux is an archaeologist specializing in cross-cultural art, ideology and environment. He has conducted fieldwork in Canada, the United States and the UK. His research ranges from ground survey and recording to theories of perception and environmental interaction in the creation and use of images and the nature of ideology. He received his PhD in archaeology at the University of Southampton.

Cultural Life at the Abyss

The Grounding of Ideology in Nature

B. L. Molyneaux

Routledge
Taylor & Francis Group
LONDON AND NEW YORK

Cover image: Atmospheres © Martha Jablonsky-Jones

First published 2023
by Routledge
4 Park Square, Milton Park, Abingdon, Oxon OX14 4RN

and by Routledge
605 Third Avenue, New York, NY 10158

Routledge is an imprint of the Taylor & Francis Group, an informa business

British Library Cataloguing-in-Publication Data
A catalogue record for this book is available from the British Library

Library of Congress Cataloging-in-Publication Data
A catalog record has been requested for this book

ISBN: 978-1-138-48407-8 (hbk)
ISBN: 978-1-138-48408-5 (pbk)
ISBN: 978-1-351-05310-5 (ebk)

DOI: 10.4324/9781351053105

Typeset in Times New Roman
by Taylor & Francis Books

Contents

Figures

Preface

The central flow of research and ideas needs the margins to keep it continuous and vital, and this is the role of outsiders. This book, a study of art and archaeology, is a call from these outlands for renewed attention to the physical systems we once identified with nature. Its goal is to help end the dominion of ideology in research about society in general and art and environmental studies in particular so that we may better understand our own integration in a physical world. I use the word "nature", which has been cast out of scholarship, because whatever one chooses to call life outside the totalizing vision of culture, it requires our careful, unmitigated attention.

The attempts I make to ground ideology in a world that humans share with other creatures, forces and things are intentionally critical of mass assumptions and normative approaches and their terminology, and so I expect that my own assumptions, intellectual habits and terminology will be criticized, and errors and inadequacies will be found. In spite of this, I want to encourage new paths of inquiry through the low cloud of relativism and representation that has settled over discourse over the past several decades.

This book did not emerge entirely in an institutional setting. My archaeological research has taken me into a variety of other circumstances, and this experience has influenced much of the ideas and attitudes I express here. After years of working and living in small communities among people of cultures different from my own, I have begun to see local expressions as the reality of cultural life. Ordinary people everywhere always struggle with remote, authoritative representations imposed from within and outside their societies that do not fit their direct experience. I have also found that a much more specific past, of actual people, places and events, better reflects the interests of these local communities. Many small societies continue to resist intense efforts over generations to deprive them of knowledge of their direct ancestors and relations in the interests of acculturation.

An interest in a past as past – as opposed to the usual idea of the past in the present – responds as well to the practical need in archaeology to understand the finality of creation. When a painting or other work of art or object is complete, it is an object of the past. Of course, objects also have futures – endless ones. We know this from the study of material culture, in which endlessly shifting meanings and functions provide a cornucopia of opportunities for interpretation. Unfortunately, these latter-day interests may also convey the idea that life is continuously recreated in every generation. Why this optimizes production in an intellectual economy is obvious, but also unfortunate, as a lack of concern with origins and formative events ignores the domain of personal experience in the life of every scholar, researcher and student. The past is immanent, in body, mind, society and experience.

I offer here two incidents that happened early in my adult life, which defined some of my future attitudes and narrowed my path through archaeology.

Long ago, living in the Northwest Territories near the Arctic Circle, I climbed in the middle of the night during the brief summer, when the sun was high, up one of the Franklin Mountains along the east side of the Mackenzie River, really just a large, rather ancient and heavily glaciated hill, to see what was on the other side. At the top and, at that point, far from a companion, I experienced no spiritual oneness, no liminal or poetic moment, the epiphany of no epiphany. In the chill winds, the sharp smells of dissolving limestone and ice in the meltwaters wafted up from the valleys and, looking northeast, I saw a vast boreal forest, extending into the haze on the horizon and the tundra of the Barrens beyond. The absence of reverie and representation was a good thing. Later on, back in the small settlement of Norman Wells, I happened to see a Sahtu Dene family I recognized heading into those same woods with a team of dogs, who pulled a fully packed sled over the dry ground, to set up a winter hunting camp.

The other was a more common occurrence, where a life's research extends from a specific event on a specific day. In my case, I happened to meet up with an artist, Selwyn Dewdney, looking for someone to help him carry on his own project, searching for and recording ancient rock paintings in northern Canada. I took this opportunity, and my own experience, and memories of Selwyn's artistic sensibility, courage and skills navigating waterways and concepts simply and directly, have stayed with me ever since.

Other events along my own crooked path, which I believe has shown some improvement, eventually led to the University of Southampton, the Theoretical Archaeology Group and the edited volume, *The Cultural Life of Images* (1997). There was a point in producing a book on images, especially related to ideas about the past, because depictions clearly took shape in the society in which they were made, and this opened up a new field of possibilities for research in visual historiography. The problem after 25 years, and much work showing how images, objects and life in the past and present are not what they seem, is that many analysts who followed did not absorb the point in the book that the problem was not with certain representations, but all representations. The work was meant to be self-critical, as even the most convincing of solutions reflects its own historical circumstances.

It was probably the timing, which was both good and bad. Postmodernism had been developing in the background for at least 20 years. In its early playful way in Italy, it poked fun at the pomposity and presumptuousness of various forms of idealism and objectivity, and disrupted and reawakened attention to issues that actual people had to deal with in society, just as Viktor Shklovsky and his formalist colleagues tried to do generations earlier in their brief fluorescence in postrevolutionary Russia. With the intensification of subjectivism into the 21st century, however, the replacement of nature with ideology showed no real change from the modernist project and, as it has throughout history, the predictable result continues to be a host of small tyrannies that promulgate instruction in how precisely to live.

In the current dominion of ideology as it appears in archaeology, there is much contestation over representations, often without consideration of the distance there is between the discourse and the material evidence of the past or, where there is a living society, the issue of imposing intellectual solutions on people who have been working out their interactions and adaptations directly over thousands of years. This is accompanied by an increasing indifference to natural systems at a time when clarity is essential.

If there is to be any new insight into the study of images in cultural life, it must address both the intellectual problem of life and the practical one of living. Change is much more difficult today. No-one anticipated how quickly virtual reality would overwhelm grounded interaction. Distance and remoteness are now banalities, everyone has a public opinion, and yet, as in the problem that Thomas Aquinas pondered in the 15th century, great ideas have less substance than can fit on the point of a pin. It seems, therefore, that the problematic in Cultural Life should have gone further than pointing out the relativism of representations, to an explicit critique of representation itself – both as an ideological issue and as a perceptual one.

It is a fond dream that there might eventually be a spiritual retrenchment that acknowledges the effects of nature on culture and, so, is more in keeping with many indigenous philosophies and the ordinary experiences of people outside the authority structures of institutions. This does not mean, however, that it is pointless to work towards such a goal. In this sense, this book is not actually about art and archaeology at all. It is my own effort to find new ways to capture worldly experience, past and present, in words.

Acknowledgements

In many ways this has been a solitary task. Art and archaeology is a modest field outside its media popularity, and the plague years have increased practical and intellectual isolation for everyone. Still, the book is done.

Key people in my educational life contributed to my point of view. Professor Ormond Mitchell, Trent University, introduced me to the writings of William Blake, and my academic supervisors Professor Joan Vastokas, Trent University, and Professor Peter J. Ucko, University of Southampton, respectively, gave me the latitude I needed to pursue my project. While my work did not follow their paths, I have always tried to live up to the examples they set for intellectual dedication, creative and critical thinking, and skepticism.

Essential material support for my research has been provided over the years by a number of institutions in Canada: the Ontario Heritage Foundation; iterations of the Ontario government's cultural ministry that governs archaeology; Trent University; the Canadian Conservation Institute; the Royal Ontario Museum, Department of New World Archaeology; Parks Canada (in particular, Rob Ferguson, who much earlier, joined me in rock art recording in the Canadian Shield and at Kejimkujik National Park); the government of Nova Scotia and the Nova Scotia Museum, who supported my crucial rock art recording during the brief draining of McGowan Lake, assisted by Helen Sheldon.

I am grateful for the institutional support I continue to receive for my research at the University of South Dakota, Department of Anthropology and Sociology, in Vermillion, South Dakota. During this period, I had the good fortune to conduct comprehensive archaeological surveys of two great national monuments for the National Park Service, Devils Tower National Monument and Mount Rushmore National Memorial (with the support of Steve DeVore, NPS Midwest Archeology Center, and Bruce Weisman, former Curator at the Memorial). This direct experience informed many of the ideas presented here. In the Department, Professor Larry Bradley's practical wisdom helped me keep more or less straight along my crooked path. Dr. Stephanie Spars' work in conflict archaeology, which looked at the acts of participants to understand the material evidence, influenced my thinking about how to ground ideology and move from empirical data into wider questions of culture and society.

I have lived and worked for almost 20 years with the Kul Wicasa Oyate, a Lakota nation and a community of people. Their values of humanity, courage, modesty, friendliness and respect have greatly influenced my life. Despite generations of harsh, forced adaptation and removals that took place as recent as the 1960s, they continue to thrive and remain constant in their goal to regain the self-sufficiency and sovereignty necessary to carry on the legacy of their ancestors and all their relatives.

Fieldwork and writing are rarely solitary in production, and a number of friends and colleagues have kept my spirits high. Dr. Nick Saunders (Emeritus Professor of Archaeology at the University of Bristol) and I were postgrads at Southampton, and that collaboration continues. Nick provided critical readings and insights in the difficult process of creating the proposal, and I got valuable ideas from Dr. Andrew Stewart, longtime friend and colleague at the Royal Ontario Museum and partner in several rock art and archaeology expeditions, and Tom Sherman (Professor in the Department of Transmedia at Syracuse University), who brought me into collaborations in art and information in the 1970s and 80s (e.g., 1 Traditional Methodology for Processing Information (Sherman 1978). Dr. Alan Costall, Emeritus Professor of Theoretical Psychology at the University of Portsmouth, and contributing author in two of my earlier books, challenged normative research in his own field, and his work in the ecology of perception is another important source of inspiration for what I have written here. I am most grateful to him for supporting my proposal and for comments thereafter. Martha Jablonsky, an artist friend from my earliest undergraduate days, graciously provided one of her strikingly beautiful paintings for the cover.

I would also like to recognize the organizations that provide free access to digital information on the internet, especially the Internet Archive (archive.org) and Project Gutenberg (www.gutenberg.org) and all the open-source communities, who provide alternatives to corporate control of information. They are essential in keeping ancient scholarship alive in a world of contemporary opinion, where most are unaware that there is little new under the sun.

Behind this book and all my writing, research and academic study is what counts in whatever path I might have taken, having love, caring and support along the way. I dedicate this all to the memory of my beloveds, my mom, dad and brother, who provided my own grounds and reasons, and, with thankfulness for their endless patience, my own family, Wendy, Freddie, Ally and Lily, who bear with me through my endless travels, long separations and preoccupations. We are now seeing in this family the next generation and the comforting sense of the flows of life. To them I offer this book, which joins a scatter of weird and wonderful writings by some of my near and remote ancestors!

1 Charting the territory

Drawing power

In his study of the forces that influence narrative, Victor Shklovsky wrote about the subtle effects of nature on creativity in a way that may help in understanding the past (2007, 26–27).

> I am like a guide through a museum or a city, who explains why the streets curve. There used to be a wall here, now we have gates, hence the curve of the street. There was a bend behind the wall, too, but that's because of the hill.

This book is an archaeological study of other places and things we encounter that also seem to have a power of their own.

High up in the dry pine forests of the Black Hills, an ancient uplift in the Great Plains of South Dakota, a spring-fed stream plunges downslope through a steep gulch past the sites of two small and unremarkable archaeological finds, located just below the springhead. One was the badly broken midsection of a black chert projectile point, which lay in pine needle duff on an eroding rock shelf several metres above the creek. Its collateral flaking suggested that it dated to near the end of the Pleistocene, about 10,000 years ago. The other was a scatter of boards, glass and barrel hoops, the remains of a still used to make illegal whisky, probably during Prohibition in the 1920s.

Archaeology separates them by millennia, content that the hunter and crafter were as different as their cultures, but these sites are little more than a stone's throw apart, and so their separate intentions seem to converge. The hunter most likely hid on a ledge over the creek, waiting for an animal grazing at the spring to pass by; and a short walk and millennia downstream, the whisky-maker hid his small operation wisely out of sight. Their bond seems to have more to do with this deep cut in the mountain slope than coincidental cultural choices.

To visualize this relationship, imagine that it happened outside time, as if thousands of years were only moments to their senses and surroundings. On that hypothetical day, after climbing down from the parched forest to the

DOI: 10.4324/9781351053105-1

stream, they looked for, and eventually found, the shelter and resources that the gulch happened to afford. And they were not alone. The hunter was probably after a Pleistocene herbivore, such as *Bison antiquus*, or an elk or deer of species that still roam the region. These animals would follow the same course along the gulch, grazing the vegetation that the stream also sheltered and watered, as their descendants still do. The archaeologists who followed in 2006 were also drawn to the shelter (Molyneaux et al. 2008). Its cooler air and water were a welcome and necessary relief from the sun, and it was natural to assume that others found this as well in the past. And finally, we must not forget the host of this event, the gulch itself, an environment with features and resources suitable for a foraging animal, hunter, whisky-maker and archaeologist to meet outside time.

Now imagine another kind of place, one with a painting or carving or other form of image or construction. Such art has an irresistible attraction, drawing attention through its forms, colours and subjects and the promise of meaning. Unlike a topographical feature, however, the integrated forces of nature and culture seem to be absent. After more than 2000 years of consuming interest, it is said that art is what it signifies, and how it fits into a specific society or a cultural process. Even when the image is on a rock outcrop or the wall of a building, what counts is almost always the picture or a feature imagined within it, despite its place in a larger environment.

The contrary idea, that an image is a feature of a physical world rather than a world in itself, becomes abundantly clear with just a short climb further up the gulch. Gradually, the drainage becomes shallower, the valley opens out and the summit of Mount Rushmore soon appears above the trees. If the hunter or the whisky-maker followed this trail to the flanks of the mountain, they would have seen a summit gnarled by millions of years of weathering into unusual animal and human shapes. In the late 19th century, an Oglala Lakota holy man named Black Elk gave it the name *Tunkasila Sakpe Paha* (Six Grandfathers Hill) (Reed and Wallace 2016, 9).

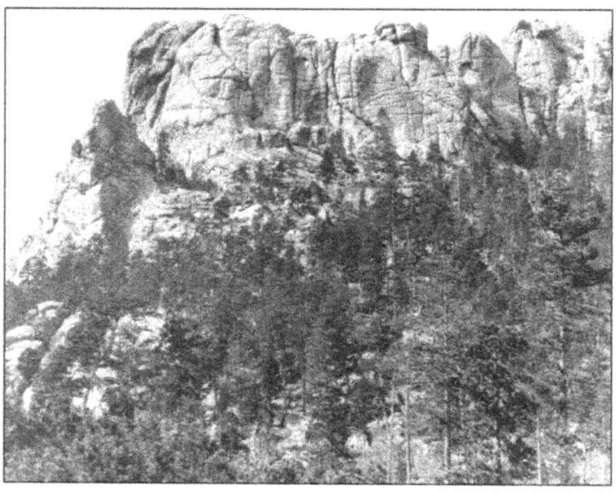

Figure 1.1 Mount Rushmore, Before 1926

By the end of the 1930s, however, these natural forms were gone.

Beginning in 1926, workers with jackhammers and explosives sculpted the southwest end into four huge faces, each around 60 feet high, representing four former presidents of the United States: Jefferson, Washington, Lincoln and Roosevelt. They look out over a huge scree slope to a vast stone and concrete plaza that is the Mount Rushmore National Memorial tourist area, and through the haze of distance, the prairies and badlands of the Great Plains.

The monopoly of attention of a monument is a form of power, and the stony gaze of each former president commands a territory with much deeper histories and traditions.

These have been the homelands and hunting grounds of indigenous people over more than 10,000 years, and they are sacred to their present inhabitants, the Lakota, yet these people are no longer free to walk or worship there. The United States acquired the Black Hills from France in 1803 as part of the Louisiana Purchase, and this brought fur traders and colonists, prospectors and settlers to the region. At first, the federal government accepted the traditional rights of several Sioux nations to the Hills, and they agreed to abide by these rights in a treaty signed at Ft. Laramie, Dakota Territory, on April 29, 1868. However, it was a hollow promise: Colonel George Armstrong Custer conducted an exploring expedition into the Hills in 1874 and found gold, and the resulting gold rush overwhelmed the ability of the Tribes to protect their territory. They continued to resist across their lands and then, in 1876, one hundred years after the Declaration of Independence, a combined force of Lakota, Cheyenne and Arapaho warriors met and wiped out Custer's Seventh Cavalry in hills above the Greasy Grass (Little Big Horn) River in eastern Montana. On February 28, 1877, Congress broke the treaty and took away the Black Hills. Since the federal government repossessed the land, they could carve on its rocks without interference. In a map of Lakota place names created by the Lower Brule Sioux Tribe (2017), the mountain is "Inyan Ihahapi", which literally means "rock laughed at, or made fun of".

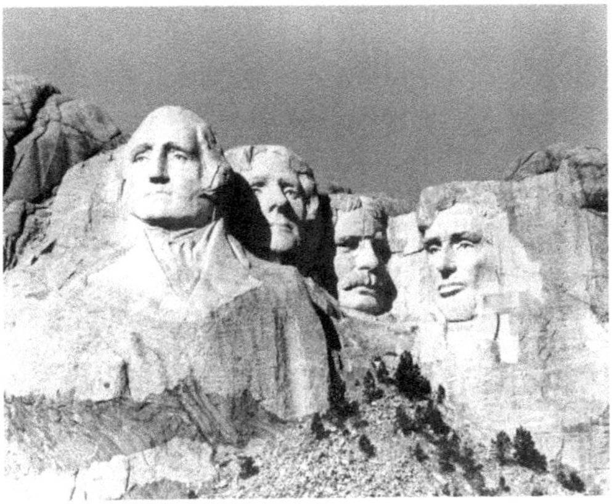

Figure 1.2 Mount Rushmore, After 1941

In 1970, three Lakota women climbed Mount Rushmore to protest the breaking of the 1868 treaty, and they were soon joined by many more supporters. The occupation lasted for months. They climbed up and camped in a small hollow behind the carvings, and draped a huge banner over the forehead of one president with the slogan "Sioux Indian Power" in red paint (Banks and Erdoes 2004, 110). They eventually left, and their struggle goes on, but material evidence of the act persists in dried fragments of the paint, scattered among pine needles on the ground (Molyneaux et al. 2008).

The result typifies the exceedingly complex power of monuments. It combines an impressive visual impact with the contrariness of a dark history and patriotic ideals of inclusive society and human equality – and it does this on profane grounds. In 1924, the State Historian for South Dakota, Doane Robinson, began searching for a sculptor who might carve some of the granite pinnacles in the core of the Black Hills to attract tourists along the state's newly built roads.

In comparison, the gulch is barely a mark on the map, but there is a telling lesson there as well. With its shelter and spring water, it has attracted and sustained diverse forms of life over thousands of years, but the small scatter of artifacts occupies and rationalizes its environment as a cultural one. It then disappears into a larger representation, a conceptual landscape of social life, in which plants grow and animals shuffle in and out of service, or do not rise to attention and, so, do not really exist at all. This human self-importance is hard to reconcile, especially next to the stream, where the meagre artifacts are lost in the rugged terrain and surrounding forest. It is, however, the microcosm of a general idea that has dominated thought over millennia – that human life is distinct from the physical world and presumes authority over its life and resources, no matter the scale.

Given this problem, it is tempting to reject culture and imagine human life as a Latourian plain, in which constantly shifting networks of people and other living things, materials and processes, dissolve hierarchies and oppositions – other than the ultimately social concerns and applications of its methods, managed through human agents and other proxies (e.g., Latour 2007). From high in the Black Hills, however, a different alternative suggests itself. It may be possible to use the idea of drawing power, rather than levelling it for the sake of equanimity in social data.

This power is not something magical that inheres in the object. Whether the feature is natural or constructed, the attraction involves the body and senses in physical relations and, therefore, extends outside the boundaries of culture. Before the powerful impact of the presidential faces and their symbolism, there was a naturally sculpted mountain with the Six Grandfathers on its summit, and before that, simply a mountain with a spring at its base, and running water that continues to carve the gulch in the slopes below.

The common recognition of natural and cultural features is a territorial bond that exists because humans, among other creatures, have evolved and continue to live through mutual relations (cf. Kropotkin 1915). Human senses, bodies and minds developed over great time like other creatures through continuous interaction with forces and materials in the physical environment, and so we experience the surroundings we have evolved to experience (Mead 1934). Thus, when provoked into attention by some striking physical feature, we may imagine, describe and interact with it as social beings, but our sensory perceptions and actions are also commensurate with those of other beings in a much larger world.

While we are only able to characterize the lure of the gulch and the mountain through the artifacts, images and beliefs they have provoked, the convergence of these behaviours and expressions over time, location and society in other places suggests that a deeper ecological relation underlies the cultural surface – one that sometimes comes into view. Standing at the edge of the gulch, it seems that we have two choices in the endless quest to understand the past: leap directly into the metaphorical abyss of language and imagination; or, first, climb down the slope and walk along the stream as one of the creatures and things that lives there. This second choice, exploring worlds beyond culture to help in the analysis of art and other meaningful environments and expressions, is the subject of this book.

The ecological grounds of art

Ecologies concerned with art and other forms of sensory expression are only possible if there is common ground between the people who make things and the people who study them. Defining this ground with respect to the past is not a simple task – epistemology and historiography show the biases of time, place, society and individual – and so the approach here is pragmatic. As artifacts are the visible effects of human–environment interactions and human senses and kinetics extend into our hominin past, culture must be a part of nature. How could it not be so? Humans emerged within a world that was not in itself a human project.

In order to move from our traditional place in the center of a world that revolves around our needs and interests, however, we need to create some metaphorical space for the non-cultural, acultural and unacknowledged in behaviour. This is difficult when all effects are treated as cultural, rationalized over centuries in individual societies through a host of models, representations and scholarship. Culture is a concept like any other, but it represents the special nature of humans and, so, appropriates all aspects of human life, past, present and future and binds them in an endless narrative cycle that is both cumulative and self-renewing. The cognitive psychologist Jerome Bruner, for example, stated that "What creates a culture, surely, must be a 'local' capacity for accruing stories of happenings of the past into some sort of diachronic structure that permits a continuity into the present" – but this "capacity" cannot reside in the individuals who contribute these stories, as "our individual autobiographies … depend on being placed within a continuity provided by a constructed and shared social history in which we locate our Selves and our individual continuities" (Bruner 1991, 19–20). Nothing can exist outside this closed circle.

The pragmatic alternative, which we will explore here, is to ground interpretation by avoiding governing concepts and narratives that organize the world and, instead, use the world humans experience directly as its own model. Whether it rises to awareness or not, this environment of physical systems exists, and its energy and materials are essential to life. Even the ostensibly social actions in the gulch and on the slopes of Rushmore reveal the energy of environmental affordances.

Allowing for forces outside culture causes problems of translation for some of the words we commonly use to describe environmental relations. *Place* is an ancient word that over time has developed far from its practical roots as locality or property. *Landscape* began as the specific view of an expanse of land captured in a picture and is now a shorthand for almost anything associated with some form of area, whether in the mind or on the ground. Together, over centuries of elaboration, they are now storage spaces for meaning, within which, it is assumed, people carry on endlessly diverse social lives. These terms are so entrenched in language that it is extremely difficult to do without them, but we will try.

The same difficulty attends the study of individual objects of material culture and, in particular, pictures and other works we define as art. Walter Benjamin, in a 1935 essay about the effects of mechanical reproduction on the perception of a work of art (Benjamin 1969, 3) wrote of its aura:

> In even the most perfect reproduction, one thing is lacking: the here and now of the work of art – its unique existence in a particular place. It is this unique existence – and nothing else – that bears the mark of the history to which the work has been subject.

This seemed to acknowledge a fundamental relation among places and cultural things, but Benjamin had other ideas. As fine art was bound in an elitist social tradition, he saw mechanical reproduction as a way to free the image from the art so that it might become a political tool accessible to the masses.

From the standpoint of the meaningful product, it may be difficult to see copying as a problem. A photo of a loved one is not diminished by its medium. And in our own age of digital reproduction, it is possible to recreate auras and loved ones in shockingly realistic ways.

Benjamin's contempt for the aura and modern disinterest in its implications for origins is unfortunate, however, because there is a rightness about it. He was interested only in the work of art in its latter-day history, as a product. He assumed that the origin of art was in ritual, and as art has always been attended by ritual, removing the aura by reproduction frees the art from this "parasitic dependence" (1969, 6).

What is missing in this and later ideological views is the sense that we have never lived in a world of objects, only works in progress from creation to consumption. All cultural things emerge in the actions of their making, and the concrete manifestations of these actions are what appear to sight.

Meanings, on the contrary, only relate to the historical afterlife of a product. While this makes for endless scholarship, as meanings vary situationally, socially and culturally, they eliminate the possibility of archaeology, and the evidence that created objects are also concrete records of their own past.

This raises a fundamental problem with the idea of representation. A painted picture, for example, is very often described as a representation, even though it is first a distribution of paint on a material surface, reflecting light, that spectators pick up from positions outside its frame. If it is more than a vehicle for the meanings attached to it, and representation is not re-presentation, as in the repetition of a sound, what can it possibly be? Here, we assume that visual perception is interaction and relation, not image projection or retrieval. As a picture is a surface with distributions of salient features encountered directly in active perception, it is not merely an intellectual object that we are socially inculcated to look at; it also qualifies as an environment within which we act (see Molyneaux 1997a).

When a picture is conceived in terms of its perception within the settings of ordinary life, the body, senses and surroundings come into play. As action does not take place in physical isolation, each sense has its own capacities and constraints, and exists in its own interactive world. Consider the differences between hearing, smell and vision. Sounds or smells radiate outwards from their source and carry through the air until they fall below the capacity of the senses to detect them. They are ubiquitous and hard to avoid. In comparison, the sight of a thing is always local and transitory. It exists only when the viewer picks it out of the available optical array, and only until they look away.

Even though vision is local and restless, images have distinct advantages over the products of other senses. Where most sounds, smells and sensations dissipate, images transmit their information as long as there is light to reflect them and people to pick them up as they pass by. Artists may therefore go to great lengths to arrange visible features in ways that attract and sustain attention. If their intent is to communicate, they may also attempt to lead the anarchical eye towards some message. Such compositions of visual features may vary over time and across cultures, and the putative messages may vary among individuals, groups or societies, but they all attract attention through their dominant features and pathways.

It still may be difficult to accept visual imagery in terms of natural forms of energy, however, because its forms strike the senses indirectly, as reflected light. As the impact of reflected light is not tangible under ordinary conditions, it may not seem to have substance at all and, thus, may seem to be something constructed in the mind. This is, of course, a major issue in studies of perception and cognition. However, as the perceptual psychologist James J. Gibson once wrote, even the most opaque symbol exists in the physical environment that we perceive as human beings (Gibson 1966, 26).

> There have to be modes of stimulation, or ways of conveying information, for any individual to perceive anything, however abstract … No

symbol exists except as it is realized in sound, projected light, mechanical contact, or the like. All knowledge rests on sensitivity.

Most importantly for research into works of art, a picture as an environment is not simply a product – a thing designed for an afterlife in material culture. The artist is a crafter and spectator, using kinetics, senses and skill in the task, and the outcome is a physical record of the acts in its creation. In this way, it is an *analogue image*. And because it is an environment, other spectators do not simply look at pictures. They occupy the completed work through their attention, scanning over the depiction and, perhaps, following the paths the artist laid out in form, texture and colour to guide the eye and the imagination.

Situations and salience

The organizing concept in an ecological study of art is the act – the situation of production. This is the past of an object or expression before its completion – what is routinely considered as its background or context. The shaping of a thing is more than preparatory work; it takes place within an environment of material and social forces – sensed or recognized – and just as the outdoor environment preserves traces of the past in its topography, the environment of a picture preserves the actions of its maker. In this, there may be features that suggest something of the maker's intent (Molyneaux 1997b).

The situation also incorporates the object into a wider field of circumstances and actions – recognizing, as in network analysis, that the relations of an object may extend far in time and space from its immediate situation. To stay close to the physical environment, we will call this material and social context, sphere of influence or interaction the deep surroundings of the locality where the production took place.

In the study of situations, the focus is on the evidence for intentional action on the part of the artist or crafter, rather than on the meanings that might accrue in what they produce. This is not the same as personal style, which characterizes the uniqueness of an individual's work over a period of time. In the depiction of a scene, for example, we might look for evidence that specific circumstances influenced the way a painter shaped an image or scene. In a study of tomb paintings during the Late Eighteenth Dynasty in Egypt (Molyneaux 1997b), images of kings seemed to vary in scale in relation to commoners as circumstances with respect to the power of the king varied in the society, as opposed to simple adherence to stylistic conventions. Or there might be evidence that an artist has specifically enhanced some feature to draw the attention of viewers. This is, of course, the general role of composition – but in a situation of production, an artist might point directly to some feature in an attempt to guide the viewer to what is most important to see. To study a picture in this way, the variable we will use to describe the drawing power of such features is a perceptual condition that crosses the barriers of time, culture and species – salience.

Salience is a condition by which a feature in an environment stands out from its surroundings (appropriately, the Latin root is *salīre*, to leap). This is the drawing power of the gulch and the mountain, and it is fundamental to active perception. All creatures navigate their environments by perceiving the salient obstacles and pathways that their senses afford.

What is most interesting about salience in the study of culture is that it has nothing to do with culture at all. Within a species, salience is a common factor in natural selection. As most living things cannot negotiate mutual interactions directly, relative salience is often crucial in attracting or directing attention in a competitive environment and, in some, relative salience seems to influence the potential for mating success. To this end, many animals and plants have sensory displays, or attractants, and these range through all the senses, depending on the way a species picks up information from its environment. Many are detectable to some extent by other species – for example, our senses pick up some of the features of the calls of birds and other animals, the sounds of insects and the colours and scents of flowering plants that are directed to others.

The human difference in this inherited biological urge is that it is often resolved in an active social intent. The philosopher John Dewey observed that salience attends meaning within the situations of daily life: "The singular object stands out conspicuously because of its especially focal and crucial position at a given time in determination of some problem of use or enjoyment which the total complex environment presents" (Dewey 1938, 67). Salience is therefore not about objects, but rather about contrasts in relations, immediate, direct and local.

Importantly for the discussion to follow, salience is physical as much as it is ideological. In ordinary experience, it may impose directly on the senses, especially when there are sharp contrasts in the field of experience, such as points of light in a night sky or a shout that turns the heads of a crowd. Such direct responses are common among humans and many other creatures. These physical aspects are, however, overwhelmed by the ubiquity of salience in situated social behaviour.

Individual people use a vast range of sensory inducements that attract personal attention, from tattoos and other body displays to ornamentation to distinctive dress, and they reach outside the body in similar ways to material culture. As social action is often bound up in social and economic and political power, appealing to the senses is a routine part of displays of status and authority, from the ornate costumes and baubles of privileged people to monumental statues and architecture about them.

When salience needs translation, knowledge and experience intervene. This falls into the domain of symbols and icons, in which a rose is *that rose*, and not simply a flower – and its study becomes the everyday problem of learning how to recognize the important features of social life. At this point, we face the abyss of meaning, and our ecological study must take it into account. It is not at all clear, however, just how much education our attention needs in these circumstances. As John Kennedy observed about Paleolithic cave art: "The modern eye can often tell what the lines and contours stand for without instruction or prompts" (1993, 269). Other animals learn about, recognize and can even make decisions without the metaphors. And, of course, making thoughtful decisions requires that individuals have the knowledge in the first place. As for the rose, training in symbolism or some form of memory imprinting to provide continuity between past and present (e.g., Olivier 2011) is probably not necessary for it to achieve its sensory effects.

Understanding salience is ultimately a question of both sense and sensibility. In a commodified world, gold and diamonds have been highly valued in many cultures over thousands of years as objects and as metaphors and symbols, and one might assume that such value relates to ideas. However, diamond is the hardest substance on earth, and gold is the metal most resistant to decay, and these material affordances are therefore part of their attraction. Salience brings attention and value to the actions of material and social systems as they meet in the confluence of human life, in the smallest of ornaments and great works of art, around spring-fed streams and on great mountains.

Managing the abyss

Dewey (1929, 24) argued that "the only way to avoid a sharp separation between the mind which is the center of the processes of experiencing and the natural world which is experienced is to acknowledge that all modes of experiencing are ways in which some genuine traits of nature come to manifest realization."

So it was back in the gulch. It developed in a flow of a spring over millennia that cut into the mountain slope, channelled the stream and afforded flows of plant and animal life through its sheltered depths. The hunter and whisky-maker who left traces there walked into it thinking that it would suit their purposes, but the gulch afforded the choices they made. Their intents were therefore inseparable from their actions, despite their preoccupations.

The discussion ahead acknowledges the inseparability of ideas and ideologies and the limited horizons of language in description, but it treats life as more than its description. In this, it does not acknowledge an ineffable god or its modern analogue, a totalizing concept of culture. Instead, it imagines that life consists in individual, direct human–environment interactions in a singular world in which culture is an aspect of nature. As Gibson observed (1979, 130):

> It is a mistake to separate the natural from the artificial as if there were two environments; artifacts have to be manufactured from natural substances. It is also a mistake to separate the cultural environment from the natural environment, as if there were a world of mental products distinct from the world of material products. There is only one world, however diverse, and all animals live in it, although we human animals have altered it to suit ourselves.

This is in keeping with language, as its metaphors must have grounds in the earth. As Deutscher argues, "we simply have no choice but to use concrete-to-abstract metaphors … if not from the physical world, where else could terms for abstracts come from?" (2005, 127). For the discussion that follows, we will gather these ideas together as the grounds of ideology.

If an epistemological issue forms the background of this approach, its motivations are practical rather than analytical, as they reflect concerns about cultural life itself. Today, there is a pressing need to redeem the idea that there is a world outside human preoccupations, narratives and representations. The problem is that we are literally losing touch with the earth.

Archaeologists see the causes clearly: the magic of tools and technologies. By this point in the 21st century, the provision of material and informational services has replaced direct production to such an extent that most people have no awareness of, or control over, the world that sustains them. And the effects are abundantly clear in studies of cultural life. In a manuscript from the 1950s, only recently published in French (2008), Maurice Merleau-Ponty writes that philosophers rarely speak about nature any more. Even the word *nature*, which once described essential or fundamental properties or qualities of the physical world, has been fantasied almost out of existence. Studies in the arts and social sciences, which ostensibly deal with human–environment relations, reflect this trend, as they often turn on ideological models and representations that may not relate back to actual sources at all. This happens in archaeology when the past and its material evidence only exist to sustain discourse, serve institutional functions and provide content for mass media.

Nature is increasingly harmed by this general inattention and increasingly harmful, as pandemics and climate change suggest. While professions of moral concern are everywhere, practical attempts to take the external world outside subjective experience into account are not so easily carried out – but that does not mean that we should simply ignore its forces and their role in cultural life.

The proposal here is to break into closed representational worlds by encouraging directness in research in material and social relations in all walks of life, including studies that depend on mind, language and ideas. The analysis draws from ecologies related to perception and action in psychology and biology, which treat life as existing within a physical world, in order to ground archaeological analysis at the scale of local populations and individual production. Work at this scale is more directly applicable to life as lived by scholars and other researchers and their subjects and may be more relevant and accessible to others in the present – when they see fit to use its records. It is motivated by the idea that many problems in history and society might better be approached directly, individual by individual and site by site, than by accumulating data and imposing solutions from above.

The discussion crosses disciplinary boundaries to test how well ideas of ecology and situation apply in different settings and with different research concerns. These discussions and case studies are arranged in a continuum of environments and situations that are increasingly distant from the literal ground – first, natural and built monuments; then, rock art, from Paleolithic paintings to late 19th century petroglyphs; next a sharp turn into the built environment, into a late medieval church with 15th century paintings; and finally, and as far as possible from the edge of our gulch, standing in a gallery and looking into a landscape painting of a bridge along the Thames in England, painted around 1840. An appendix provides direct links to high-resolution digital images of the most well-known of these works, as they provide colour and detail outside the capabilities of paper illustrations.

Some of these works, most from original fieldwork, are very well-known and heavily interpreted within their fields, and in some instances in many, and the analysis skirts over many significant issues and ideas. This is only to help define fields of possibility for future ecologies of art and other expressions of cultural life. In all these examples, the focus is on display and its salience, from exaggerated expressions of power to mild frissons more in keeping with ordinary experience, in the belief that this is the hidden world of energy showing through. But first the notion that humans have authority over nature must be expunged so that we may begin looking in this relatively uncharted territory from common ground.

References

Banks, Dennis, with Richard Erdoes. 2004. *Ojibwa Warrior. Dennis Banks and the Rise of the American Indian Movement*. Norman: University of Oklahoma Press.

Benjamin, Walter. [1935] 1969. The Work of Art in the Age of Mechanical Reproduction. In *Illuminations*, edited by Hannah Arendt, translated by Harry Zohn. From the 1935 essay. New York: Schocken Books.

Bruner, Jerome. 1991. The Narrative Construction of Reality. *Critical Inquiry* 18:1–21.

Deutscher, Guy. 2005. *The Unfolding of Language*. New York: Metropolitan Books.

Dewey, John. 1929. *Experience and Nature*. London: George Allen & Unwin.

Dewey, John. 1938. *Logic: The Theory of Inquiry*. New York: Holt, Rinehart and Winston.

Gibson, James J. 1966. *The Senses Considered as Perceptual Systems*. Boston: Houghton Mifflin.

Gibson, James J. 1979. *The Ecological Approach to Visual Perception*. Boston: Houghton Mifflin.

Kennedy, John. 1993. *Drawing and the Blind*. New Haven: Yale University Press.

Kropotkin, Petr Alekseevich. 1915. *Mutual Aid. A Factor In Evolution*. London: Heinemann.

Latour, Bruno. 2007. *Reassembling the Social: An Introduction to Actor-Network Theory*. Oxford: Oxford University Press.

Lower Brule Sioux Tribe. 2017. *Map of Lakota Place Names within the Great Sioux Nation Lands of the 1868 Fort Laramie Treaty*. Lower Brule: Lower Brule Sioux Tribe.

Mead, George H. 1934. *The Philosophy of the Act*, edited by Charles W.Morris. Chicago: University of Chicago Press.

Merleau-Ponty, Maurice. 2008. La Nature ou le Monde du Silence. In *Maurice Merleau-Ponty*, edited by Emmanuel de Saint Aubert, 44–53. Paris: Hermann Philosophie.

Molyneaux, Brian L. 1997a. Introduction: the Cultural Life of Images. In *The Cultural Life of Images: Visual Representation in Archaeology*, edited by Brian L. Molyneaux, 1–10. London: Routledge.

Molyneaux, Brian L. 1997b. Representation and reality in private tombs of the late eighteenth dynasty, Egypt: an approach to the study of the shape of meaning. In *The Cultural Life of Images: Visual Representation in Archaeology*, edited by Brian L.Molyneaux, 108–129. London: Routledge.

Molyneaux, Brian L., Stephanie A. Spars, Mandy M. Klein, Nick Schultz, and Melissa E. VanOtterloo. 2008. *A Park-Wide Cultural Resources Survey of Mount Rushmore National Memorial*, Pennington County, South Dakota: Final Report of the Class III Reconnaissance and Intensive Archaeological Surveys, 2006–2007. Permit #: MORU-2006-SCI-0002. National Park Service, Mount Rushmore National Memorial, Keystone, South Dakota.

Olivier, Laurent. 2011. *The Dark Abyss of Time: Archaeology and Memory*. London: Rowman & Littlefield.

Reed, Paula S., and Edith B. Wallace. 2016. *Shrine of Democracy and Sacred Stone: Historic Resource Study, Mount Rushmore National Memorial, South Dakota*. Washington: National Park Service.

Shklovsky, Viktor. 2007. *Energy of Delusion: A Book on Plot*. Translated by Shushan Avagyan. London: Dalkey Archive Press.

2 Mere nature

The rise of authority

In 1726, Daniel Defoe, the author of *Robinson Crusoe*, learned of the capture of a boy who had somehow grown up on his own and lived wild in a forest in Germany. The boy eventually became a protégé of George I, and this amazing sequence of events inspired Defoe to write a pamphlet dealing with questions of man and nature. He concluded that all humans were endowed with souls, but this godly possession did not confer humanity. Man required education to rise from mere nature, the background of animal and vegetal life.

> Mere Nature receives the vivifying Influence in Generation, but requires the Help of Art to bring it to Perfection of living …
>
> (Defoe 1726, 61)

Defoe's pamphlet shows the retreat of an active god into a vague life force, which was an essential condition for the rise of modern science and humanities. And it fit with John Locke's idea that humans begin as a blank slate and learn by experience, which they do by being members of a society ([1690] 1721). However, the divergence of being into educated humanity and mere nature contradicted the traditional belief that life and matter were integrated in a single reality, however magical its guiding forces were imagined to be. In doing so, this new attitude eventually caused a divergence in research about human life, with the educated arts going one way, and the sciences of mere nature another.

There are good reasons to reunite them. For all practical purposes, physical nature may be ineffable, but its energy resources and transformations still afford and constrain the possibilities of life. One possible way to bring them closer together is to accept in a practical way that forces exist outside culture. Accepting a material "other" seems preferable to the current state of humanism, in which revolving debates create their own intellectual objects and subjects of study without reference to their physical dynamics. These humanist projects miss a great opportunity, as they ignore the working world of cultural life – the mutual interactions of people in their material and social surroundings, which enable the human project to exist in the first place.

DOI: 10.4324/9781351053105-2

Unfortunately, the idea that humans inhabit a special world is an ancient phenomenon, and so deeply embedded in our concepts that it is almost impossible to think otherwise. Indeed, a student of culture learns early on that the culture concept is elusive, with as many definitions as people who think about it. There is a consensus, however, about what it is not – nature. This conceptual distancing continues to encourage the traditional idea that humans have authority over nature – and, thus, authority over the actions and processes of life.

This attitude probably began in the early Holocene, in a technological shift from strategic foraging, which involves plant and animal management, to animal husbandry and horticulture, in which people do some of the active work of creation. It would not be an ancient revolution, but rather a shift in attention from the affordances of nature, the primary resources that simply exist where they are found, to the secondary resources produced in more permanent settlements. As people spent more time involved in a world they created and transformed than in the raw environment, they invested more thought and meaning there. In the area of our interest, for example, many hunting and gathering societies in the ancient past had a special regard for conspicuous natural forms, as rock art and oral traditions suggest, but only horticultural societies had the ways and means in people, technology and resources to build places for ideology to reside and afford their specific social intents.

The housing of information introduced a new problem. In a practical sense, inhabitants of sedentary societies are continuously immersed in social intents – the settlements in which they live – and so they are much more vulnerable to ideological control than mobile societies surrounded by an indifferent nature, where most of the dangers are physical. When hunter-gatherers, in the rhythms of their ecology, cleaved to places to which they attributed great power, they did so in a larger world they did not control, whereas more sedentary horticulturalists saw their own constructions rightly as worlds of their own devising and so could claim some of the power as their own.

Visible demonstrations of authority over nature appear first in the ancient civilizations of west and south Asia, in stories of gods causing great deluges and floods that renewed a decayed and deplorable earth for the benefit of human beings. In the earliest of three Babylonian accounts, dated to around 1800 BCE (Lambert and Millard 1969), the god Enki warns the hero Atrahasis to dismantle his house and build a boat to escape the flood planned by the god Enlil to destroy humankind; Atrahasis boards the boat with his family and animals seals the door, and the flood swells for seven days; and at the end, Atrahasis offers sacrifices to the gods, and his efforts replenish the world. The story of Noah, in the first book of the Bible, emphasizes the separateness of humankind and its dominion over nature:

> the dread of you shall be upon every beast of the earth, and upon every fowl of the air, upon all that moveth upon the earth, and upon all the fishes of the sea; into your hand are they delivered.
>
> (Genesis 8:3 King James Version (KJV))

The wide distribution of such ideas is evident in the Matsya Purana, a Sanscrit text dated to the 3rd century AD, which has a much different hagiography, but shares the theme of cleansing and renewal (Dimmitt et al. 1978).

These flood stories are millenarian solutions to issues in ordinary life, and so we should treat them as theoretical, philosophical or intellectual positions in relation to specific social problems of the time, rather than mythologies conjured up in some state of innocence. Evidence for a more pragmatic view of mythologies of nature and society is especially clear in the original human crisis in the Bible, Torah and Quran, the fall of Adam and Eve. As the story goes in the Bible (Genesis 2:4–3:24 KJV), God forbids Adam and Eve from eating the fruit of a tree standing at the center of the Garden of Eden; a cunning serpent convinces Eve to eat of this fruit and share it with Adam; the couple at once find themselves naked and so hide in the garden; and then, when God discovers what they have done, he banishes them to a harsh, mortal life in the deserts beyond Eden.

This seems like a parable associated with the risks of disobedience to authority. Yet there was something special about the tree, which justified its central place in the garden. When Adam and Eve eat its forbidden fruit, God says: "man is become as one of us, to know good and evil" (Genesis 3:22 KJV), and so he casts them out before they eat from the tree of life and become immortal. It seems that the great sin, and the threat in all social hierarchies past and present, is awareness.

Allusions to authority in society became graphic during the fluorescence of philosophy and art in Greece in the fifth and fourth centuries BCE, during which some scholars worked specifically on problems of being. Aristotle (384–322 BCE) organized being in a *scala naturae*, a natural ladder, set on the inanimate earth, rising through the plants and animals to humans at the top: "Nature passes so gradually from inanimate to animate things, that from their continuity their boundary and the mean between them is indistinct" (Cresswell 1887, 8.1.2). His hierarchy acknowledged that while some mysterious life force animated all living things, humans had more complex capacities than other beings. They could act in ways that seemed closer to animals (raw emotions and desires), but also think about actions wilfully and, ideally, with reason (morals and intellect). Aristotle's ladder was therefore a practical observation of dynamics in the world around him.

After the decline of Athens, as Roman civilization spread and in the 3rd century AD Christianity extended over Europe, the order of authority reversed. Following the hierarchy of the Bible, being became a chain that began in the heavens, with God on high, surrounded by hosts of angels, and then descended into the world as Adam and Eve fell, with declining authority and sensibility, to other humans, animals, plants and finally to the lifeless earth (cf. Lovejoy 1936). Of course, this dystopian version of human nature was equally informed and pragmatic. Its logic fitted the masses of ordinary people at the bottom, landless and illiterate, who did not get much benefit from God's earthly bounty. For these masses, ordinary life – the day-to-day task of maintaining one's existence – took place among the other creatures of the earth. In a supreme rationale, however, the great chain presented the promise of upward redemption through dutiful moral education – just as Defoe would argue for higher society generations later.

These ancient expressions of authority over nature suggest that the distinction emerged in concerns with ideology and social control. The lowly place afforded to nature is quite ironic, however, given the appeal to the base senses of ritual and ceremony in churches and temples. Religious architecture is usually devoted to sanctuary, the creation of a peaceful and thankful retreat from natural life, with only enough window space to let light in and keep nature out. Yet rituals in all but the most ascetic religions are redolent with sensory appeal – orations of scripture, chants, song, music, colourful and striking costume, choreographed movements and processions, ornate furniture and art. Clearly, priests and congregations needed the stimulation of the coarser earth to engage belief.

In western Europe, religious authority began to decline in the 15th century, as a merchant class developed, and nature once again became a tool for the intellect. The impetus came as traders encountered vastly more diversity in human and animal life than in the unified, highly integrated world presented in the Bible. Signs of this change are everywhere in the intellectual movements of the Renaissance and the Enlightenment, but most of the attention was on the new horizons of human beings.

However, with respect to nature, one current that remained consistently focused on the relation between humans and other creatures, which Defoe would have found quite curious, was the study of anatomy. In this field, the differences between humans and animals were not so well defined, and when the issue arose, it was worked out in dissection.

In the mythologies of hunting and gathering cultures, supernatural beings were often able to change shape at will and, in many, humans and animals had parallel societies. This fluidity did not come only from the imagination; most families hunted and butchered their own animals for food. For example, when the Ojibwa in northern Ontario speak of a bearwalker, a shaman who can take on the appearance of a bear (e.g., Warren 1984, 110), they take it not from religious reverie, but from direct observation and experience. Bears often walk on two legs and sometimes sit in a remarkably human way, and when a bear is killed and skinned, its resemblance is even more striking, especially so in the hands. In the United States, the Fish and Wildlife Service published a manual for forensic scientists and police forces to help distinguish between the two (Sims 2007).

In Europe, such resemblances were acceptable when there was still magic in the world. Galen (129 – ca. 210 CE), the Greek philosopher and physician who wrote the first treatises on human anatomy, was forbidden to use human cadavers, and so he dissected Barbary apes and pigs because of their physical similarities (Singer 1956). And as long as the Church had its authority, the official position on dissection was inconsistent (see Ghosh 2015). Yet, for all the research in human anatomy in these centuries, Galen's treatise on animal subjects, after more than a thousand years, remained the primary source for human anatomists until Vesalius published *De humani corporis fabrica libri septem* (Vesalius 1543). This complacency suggests that the clear distinction of humans from animals was simply not an issue. After all, the lives of humans and animals were intertwined in many aspects of ordinary life, from agriculture and food to transportation to all the metaphorical relations in oral traditions and popular literature. By the time of Vesalius, scholars also had more pressing concerns. In the same year that the *Fabrica* appeared, Copernicus published his book on the revolutions of the planets, which showed that the sun, not the earth, was the center of the cosmos (Copernicus 1543). This shift away from God and the earth, combined with the burgeoning preoccupation with all things human, began to dissolve the mythological unity expressed for thousands of years in the Garden of Eden.

While increasingly secularized scholars either worried about or celebrated humanity, however, interest in the relationship of humans and animals continued in the background – even Descartes noted in his *Discourse on Method* (1637) that the hearts of humans and large animals were comparable (Cottingham et al. 1985). These interests became explicit in the work of the English anatomist Edward Tyson, who began to publish dissections of animals and compared them to humans. Among his comparisons was one that anticipated things to come, when he concluded that the brains of chimpanzees looked more like human brains than they did the brains of monkeys (Tyson 1699).

Thus, by the time Defoe confronted the problem of the feral boy in 1726, angels had retreated and a less certain god conferred a soul in humans, and the main source of humanity was society – that is, for those who had the opportunity for education. Peasants and indigenous peoples lived closer to mere nature, which became a catchment for outsiders, animals and the rest of the natural world, their study consigned to natural history and, eventually, the sciences.

Just nine years after Defoe published his pamphlet, in 1735, Carolus Linnaeus published his first edition of a taxonomy of living things, *Systema Naturae*, which would soon end scientific notions of the singularity of human beings, and it expanded quickly in many editions (e.g., Linnaei 1758). He revealed that under God's creation, humans were members of Homo, and Homo was a member of the Anthropomorpha (later to become Primates) along with Simia (apes and monkeys) and Bradypus (sloths). And a little less than a hundred years after this, two works completed the idea of a unified science, if not the scholarly world. Charles Lyell (1830) showed that geological processes were uniform across the world – and then Darwin (1859) affirmed a similar unity for biology, confirming the Linnaean descent of humans from other primates.

A new type of culture

It might seem surprising that these profound discoveries did not encourage more integration between human beings and the rest of the world. Even if humans did evolve from other primates, however, the humanities had an epistemology based on the Bible and the hermeneutic traditions of its scholars. As a result, the discoveries of indigenous hunting and gathering societies encouraged the idea that human life moved in a stable and orderly progression upwards, from an original, primitive condition. Thus, when European merchants and explorers began to dominate the sea-going trade routes in the search for resources, and encountered people first-hand who did not fit within their ideas of society, they could place them in an orderly world.

At best, these foragers, hunters and horticulturalists seemed to be part of an Adamic world before the Fall, communing with animals, worshipping them and seeing spirits in virtually every form of matter. At worst, they were subhuman or non-human beasts and devils. In this atmosphere, evolution was irresistible as a solution. To western eyes, the difference between simple and complex societies related to progress, the logic of the great chain of being without god and the angels. In a twist of semantics back to Aristotle's ladder, humans ascended, rather than descended, from apes (cf. Darwin 1871).

Edward Tylor first redefined the notion of culture that was common in Defoe's time (1871, 1):

> Culture or Civilization, taken in its wide ethnographic sense, is that complex whole which includes knowledge, belief, art, morals, law, custom, and any other (capabilities and habits) acquired by man as a member of society.

This new kind of culture united people of all kinds into a single humanity, from hunters and gatherers to the industrial nations of the world, and because it drew from Darwin's integration of all creatures into one system, it applied, like Lyell's uniformitarianism, at all scales of human endeavour, from behaviour to the patterns of social groups, and all areas of enquiry.

By the mid-20th century, anthropology had a universal reach. With its comparative approach, it provided a means to explore the great diversity of cultures and societies across the world in a single vision of humanity, of course with much debate about the theories and methods in doing so. Even then, there was still a sense of difference between humans and nature, as is evident in this definition by Leslie White 1949, 1), which is reminiscent of Defoe:

> Every individual of the human species is born into a cultural environment as well as a natural one. And the culture into which he is born embraces him and conditions his behavior.

From this point on, however, the distinction between nature and culture began to erode. By the 1960s, there was no need for gods or millenarian beliefs to demonstrate authority over nature. Human life was naturalized in the idea of culture, and culture largely replaced God in dominion over the earth. Every human pursuit was a cultural one, whether the focus was an individual, a social group or a particular way of life. As for what might be unappropriated – undiscovered people among their undescribed plants, animals and natural materials – it was only a matter of time before culture would take them in.

Meaning as totality

The freeing of humanism from the constraints of empiricism enabled researchers to deal more easily with the ideological aspects of life, as they could confront the abyss with metaphors, but this approach was fragile in an unexpected way.

Among all these developments in anthropology, there was a tacit understanding that culture exists as an entity – carefully described as an adaptation – outside the specific groups it organized and its processes governed. People were people because they had culture. This new version of the collective metaphor satisfied what was obvious to anyone who traveled outside familiar environments; individuals, extended families and communities could be distinguished from others by their social traits and also organized into larger societies and cultures. Anthropology in its various forms thus had the licence to expose the rich tapestry of human life and, for some, the invisible cultural processes that made it all work together.

However, early 20th century discoveries of fundamental indeterminacy in physics and mathematics eventually worked their way into the social sciences and humanities. Thus began an epistemological shift towards something the literary imagination had assumed all along – that objectivity was not possible. And if objectivity was not possible, neither were assumptions of uniformity in human development, progress, social and cultural boundaries and, eventually, the reliability of observations and descriptions outside individuals.

Ironically, however, indeterminacy actually enhanced the determinacy of the culture concept. When the humanities eventually began to see science as a social pursuit, culture became free of all material constraints and replaced even the qualified nature of science as the arbiter of all human life. By the turn of the 21st century, the result was inevitable. After centuries of humanism, cultural studies especially seemed to revolve helplessly in a closed world (Layton and Ucko 1999, 12), which valorized the primacy of mind and, inside its discourse, subjective anarchy and unfettered imagination.

Culture as a totalizing concept resembles a universalizing religion. It exists everywhere that meaning exists, from individual bodies to the frontiers of the imagination. New ground or novel experience is instantly colonized within the horizons of language. And there is a sociology or anthropology of everything. As in ancient visions of the cosmos, reality revolves around the human subject, perpetuating the biblical mandate for humans to have dominion over the rest of the earth. And as in the days of the ancient biblical scholars and their authorities, cultural life is a matter of interpretation – through discourse, tacit agreement within intellectual paradigms or the pressure of public opinion.

This state of affairs is difficult to grasp for any archaeologist who tries to understand the past life of material things. It is not that physical processes such as kinetic energy are simply absorbed into human intentions – as if culture propels the hunter's arrow or a potter's foot turning a wheel is the force spinning the lump of clay outwards in the hands of a potter shaping a pot. The point is that it does not matter. With the analytical turn increasingly towards the putative meaning of things, even the senses are interpreted as social acts and ideological tools and, like the gaze in cultural theory, they become powerful metaphors in the study of social interaction. We may owe a debt to science for physiological and neurological studies, it seems, but it is not necessary in archaeology and other studies of culture (e.g., Day 2013, 2).

The impossible past

The rise of the meaningful subject severely tested archaeologists at the beginning of the 21st century, already juggling with humanities and sciences in the attempt to make sense of the past from material things. Once a clue to what happened in history, the artifact was increasingly an ephemeral object, an anchor or node in personal and social discourse, replete with putative meanings available to be contested or negotiated. It seemed impossible to interpret the past except in terms of the present.

This impasse encouraged a drift from empirical approaches in university settings towards interest in individual and social experience, and concepts such as material culture and cultural landscapes became ways to offer narrative solutions to the problems of the past. Because there were no empirical constraints, this new way of seeing dramatically increased the interpretive potential of artifacts, as they become things with cultural lives (e.g., the papers in Molyneaux 1997a).

The shift to narratives about the meanings and functions of artifacts has been highly beneficial to archaeology as an intellectual pursuit. The latter-day interpretation of material culture is never short of things or ideas. For field archaeologists, however, the experience has been strikingly different. They could once present a general sense of prehistoric life, which they drew from artifacts and materials recorded in surveys and excavations; and from this evidence, they could sort the inhabitants of a region into social and cultural groups and show some of the ordinary dynamics of social life – interaction, change, movement and migration. However, because this research essentially refers to a past that is fundamentally archaeological, it lacks the ideological richness of living cultures, especially those of the descendants of the people they study – and these nations and social groups have no need for archaeology and can interpret their own pasts as they see fit.

Archaeology in general has therefore become more ideological than empirical, with exploration, research methods and results often negotiated in order to harmonize with sociopolitical realities. In this pragmatic world, archaeological purism tends to lose the argument, as its intellectual concerns and result may be irrelevant, intrusive or harmful to people within their own horizons of belief. As a result, much of the material evidence of the past is now inaccessible, and archaeology increasingly resembles anthropology or cultural studies, presiding over an empire of ideas resolved in discourse.

The groundless present

The concept of culture has achieved what religion could never do, which is remove people, ideologies and analytics from the sense that they are part of the nature of all things, the living materials of the earth and all its creatures.

Marshall McLuhan (1962) famously predicted that advances in communications technology would transform the world into a global village and, now, with a global internet and social networking, people are able to acquire information and services and interact with others without any reference to the outside world or to people in the flesh. The active world is now more than ever experienced through these simulacra, which require a very narrow range of physical and sensory actions to process.

As direct, physical encounters continue to decline in the shift from a grounded reality to placeless digital experience, cultural life exists mainly in hierarchical networks of computers and smartphones, nurtured by corporations and sustained in their servers. In this virtual web, image and text are replacing physical nature as the sources of cultural life. In post-industrial societies, more and more urban people work and receive virtually all services and supplies directly or remotely in their workplaces and homes, spending much more than 90 percent of life indoors – an observation made in a research study 25 years ago (Tsang and Klepies 1996) and still relevant. And when people do go outdoors, they typically experience only managed versions of the outside world, passive arrangements of plants, in gardens, parks and waste spaces.

This is not McLuhan's village. In another prophetic statement he made in 1964, he said that "Leasing our eyes and ears and nerves to commercial interests is like handing over the common speech to a private corporation, or like giving the earth's atmosphere to a company as a monopoly" (1964, 15). And here we are in the 21st century, having done just that!

It is not surprising in this interior atmosphere that the routine metaphorical jargon of theoretical analysis derives from this urban economic infrastructure. Theory building, packaging and unpacking, constructing and deconstructing suggest that analysts think they are the observers and inhabitants of discourse, independent of the physical world of their creations. In comparison, the metaphors of nature, from ancient notions of mysterious powers to the wider communities of life, seem quaint and primitive. Nature is now a question of semantics, a description or metaphor, a commodity or brand, or nothing. And continued technological development will only intensify this separation. Old social hegemonies, families, religions and societies are increasingly replaced by corporations and other complex institutions, which increasingly control social production, social networking and the propagation and nurture of ideas. Guy Debord wrote of this problem in 1967, long before smartphones were in every hand (1995, 24):

> Though separated from his product, man is more and more, and ever more powerfully, the producer of every detail of his world. The closer his life comes to being his own creation, the more drastically is he cut off from that life.

While it seems that service industries and institutions have replaced the ecologies of nature, even in this dystopian scenario, the dynamic interactions with nature are merely ignored. This ordinary nature is in everyday experience. As I type this *sub rosa*, each keystroke, rendered with Boolean precision, travels through an architecture of electrical circuits before it appears in an approved font on the 3840 x 2160 pixel screen. At the same time, the keys click and I see my reflection in a beam of light hitting the laptop, while wind, birds, barking dogs and passing cars compete outside my window. As nature inevitably shows itself in even the most disembodied tasks, it makes sense to shift some attention to the unanticipated, unacknowledged perceptions and actions of physical experience – Merleau-Ponty's (2008) "brute nature", which still evades appropriation.

His concern was justified, as modernism and its presumptions of order gave way to a disorderly postmodernism that questioned authorities and standards on all fronts. Of course, this was all in the mind – a matter of choice as to how to manifest ideas about life. The architect Robert Venturi, for example, wrote in 1966 that he was "for richness of meaning rather than clarity of meaning; for the implicit function as well as the explicit function" (1977, 16). However, while it seemed to some that this approach better suited a world that seemed to have no unity, postmodernism only roiled the intellectual economy. For all the subsequent playfulness with architectural form, postmodern buildings still needed the old unities of physics and engineering, even as they are scarred by indeterminacy, and the forces of physical nature continue to be as present in the homes, offices and streets of scholars of even the most radical of urban persuasions, just as they are in the remotest settlements of the earth.

Perhaps the problem in our inability to express the obviousness of forces in nature is not a problem of nature but a problem of culture. As a totalizing concept, culture does not allow consideration of human life outside its representations – the outlands where brute, wild or mere nature enables, constrains and affords existence. Latour once described this outland as "mere nature" (Latour 1993, 137) just as Defoe before him, but only as an extreme to a transcendental human that could never exist. In place of such polarities between discourse and nature, he imagined a parliament of things, with humanists and scientists reunited in their sociopolitical contexts, negotiating their mutual self-interests. As archaeology works in a world of silence, it has to look another way, not mediate, negotiate or contest meanings, but rather search for comparable grounds.

Comparable natures

Humans do not live in representations. Like other living things, we interact in the flows of material and social forces we encounter in daily life, a fundamental relation we share with many other creatures. As Darwin observed (1871, 186):

> Some naturalists, from being deeply impressed with the mental and spiritual powers of man, have divided the whole organic world into three kingdoms, the Human, the Animal, and the Vegetable, thus giving to man a separate kingdom. Spiritual powers cannot be compared or classed by the naturalist; but he may endeavour to shew, as I have done, that the mental faculties of man and the lower animals do not differ in kind, although immensely in degree.

What is important about this integrative point of view relates to the imagined boundaries of humanity. Whether conferred by a god and managed in a ladder or chain of being or a Darwinian evolutionary tree, human physical systems developed long before the relatively late emergence of the intellect – and this means that sensory and kinetic activity retain these adaptive features. This suggests that to understand cultural life, we have to confront natural life – and that means trying to find ways of seeing outside the horizons of language and society.

Tim Ingold once wrote about the web spider, and how the threads of a web were "extensions of the spider's very being as it trails into the environment" (2010, 210–11), recognizing that as the web and web spider are, in a way, indivisible, so are we from our own environments.

There are other ways of seeing spiders. Among the Lakota, for example, Iktomi is a spider-trickster, whose exploits, especially with Coyote, show the contradictory and unpredictable forces of all life (Zitkala-Sa 1985). With this in mind, consider the creature that hunts the web spider and so does harm to this unifying representation.

Salticidae, or jumping spiders, are roving, predatory insects with highly evolved visual systems. They have two sets of eyes, one for visual acuity, and one for motion detection, and as this acuity reaches almost human levels, these creatures are masters of their environments. They also seem to be able to integrate the inputs from these eyes in a non-linear way, as if incorporating choice (Menda et al. 2014). This hint of intentionality is clearly exhibited in salticid hunting behaviour. In one tactic for catching a favourite prey, web spiders, a salticid vibrates the web with its legs and palps as if it were caught, and then leaps on the unsuspecting victim as it rushes out to investigate (Richman and Jackson 1992).

Some Portia species exhibit features ordinarily associated with learned behaviour, assumed to be restricted to more complex species. As this spider nears its prey, it may adopt cat-like stalking behaviour and, most surprising, may choose to move from the line of sight and detour around an obstacle instead to avoid detection until the last moment (Harland and Jackson 2000). To accomplish this feat, it must somehow internalize aspects of the environment and recall its goal (Tarsitano and Jackson 1997). Otherwise, it would simply wander off. Also typical of more complex animals, many jumping spiders have large repertoires of often complex displays when a male and female come into contact before mating (e.g., Jackson and Harding, 1982).

We cannot imagine what a jumping spider might experience as it moves about the environment, stalks and deceives prey, defends its territory and attracts mates with display – but we do know that it does all these things with a brain the size of a poppy seed (Menda et al. 2014, 1). If such a tiny creature can accomplish tasks so familiar to humans that it lures us into anthropomorphism, perhaps it is because so much of being human, mind and body, also extends deeply into collective life on earth. We are not above the organism–environment interactions of other creatures, even though we may characterize this forceful ground of activity as our own, in metaphorical descriptions and interpretations.

To resist drawing human life into web-like representations centered on a cultural brain and social matrix, which only maintain the golden divide and the special nature of humanity, we need to identify commensurate features among humans and other creatures that are consistent with our slow emergence and adaptations over time.

We humans are a social species with an abiding interest in ourselves and the ability to make choices as to how to live, but we do so with bodies and senses that are the result of millions of years of adaptation to a relatively narrow range of environments. This confluence of body, acultural mind and environment constitutes the grounds of culture.

Jakob von Uexküll ([1934] 1957) expressed this idea in the concept of the *umwelt*, which emphasizes that the environment that each organism experiences is relative to its scale and sensory systems and the ways and means of its transformations of energy. While the umwelt accounts for generally uniform patterns of behaviour within species, localized physical processes are not uniform in effect. Just as water finds its own path down slopes and around obstacles, so the flows of human life vary with local surroundings.

It might seem that the umwelt does not quite fit within the human project. Humans have overcome so many of nature's constraints through invention and engineering, and somehow gained self-awareness in the process, that home, community and society seem to have replaced nature at the center of the world. Even within social horizons, however, our territorial bond still holds true. Humans are not isolates, despite the imaginative boundaries we have set over time. In Mead's account of this process, life is not something that occurs within the organism; it is a dynamic material relation in which the actions of the environment alter the organism, and the actions of the organism alter the environment (Mead 1934b).

These integrative views of life suggest that nature is not a thing acted upon by cultural beings, or an autonomous force, or machinery that hums in the background. Cultural acts, events and processes, picked out in analysis from ordinary life as it happens, may overwhelm, but do not replace the physical experience of humans as fellow beings in a larger physical world (see Costall 1985). As Mead writes, each organism selects its environment "by its own sensitivity" (Mead 1934a, 132).

These wider connections are well recognized in indigenous societies – and it is not because of a simple anthropomorphism of other creatures for the sake of mythology. They recognize common behaviours through their lifelong, intense interactions with the plant and animal life and surrounding environment.

Perception as an ecology

The perceived world is traditionally seen in terms of representations and other forms of illusion produced in the brain (Costall 2007). However, if human action is a recursive relationship with the environment, the perceived world cannot reside somewhere in the brain, as a backdrop for meaningful action, a picture in the mind or an isolated outpost that exists only when lit by some searchlight of awareness.

The alternative is the theory that perception is an ecological activity, an idea that the perceptual psychologist James J. Gibson developed in the mid-20th century (1966; 1979). Gibson was concerned with practical vision. In common with Dewey and Mead, he rejected the traditional Cartesian distinction between organisms and environment (see Costall 1984). He proposed that as perception is an interaction between a perceiver and the optical array available in the immediate surroundings, the light that reaches the eye must contain visual information already structured by its properties and features, as opposed to being resolved in the brain. This has great significance for the study of human behaviour. If perception is direct, humans live *within* environments, rather than presiding over them, like observers of nature.

To describe how individuals choose to navigate within perceived space, Gibson introduced the concept of affordances, which he defined as "opportunities for perception, of available information, of potential stimuli" (Gibson 1966, 23). In Costall's lucid example, "we perceive that a surface affords walking in the very activity" (1986, 46). This exposes a crucial difference from traditional representationalist views. Affordances are not the features of a conceptual landscape that individuals carry about in memory, like the components of a mental map. If perception is direct, there is no *need* for visualization, and thus no need to invoke governing processes, representations or plans in the study of cultural life.

Most neurological research supports representationalist approaches (Costall 1984), but some recent discoveries suggest that the brain may participate more directly in navigation.

In the 1970s, the neurologist John O'Keefe discovered that when a rat moved along a set course, certain neurons in the hippocampus repeatedly fired when it passed by specific features, thereby charting movement outside awareness and without the intervention of other processes (O'Keefe and Dostrovsky 1971; O'Keefe and Nadel 1978). As Louise Barrett summarizes:

> neurons that are next to each other will fire in sequence when a stimulus moves across adjacent positions in the sensory field, so that one can often recognize what the organism is viewing from the pattern of neural activation produced.
>
> (Barrett 2011, 146)

Continuing research on place cells in the human hippocampus suggests that the distributed work of the active individual, senses and brain orient the body in the environment prior to awareness – that is, without the intervention of conscious recollection, representation or, in some as yet undefined background, culture.

Ecological perception is important here because it provides support for two aspects of an ecology of art and its situational approach. If perception and action are direct, behaviour is not encompassed by the distractions and preoccupations that social awareness brings to daily life. It also means that there is no logic in the golden divide that separates humans and animals. All living things have emerged and adapted through active perception in the environment available through the senses, and so, while knowledge complicates an individual's experience of the world, the physical processes involved in sensory activity are common to all – including the dynamics of display. Because the human–animal divide is so fully institutionalized, a more integrated study of perceptual relations has yet to emerge. It was only in 1992 that Dusenbery (1992) established the study of sensory ecology, which deals with how organisms sense and use the environments they experience and, in practice, this field remains mainly an ethological one.

Ecological approaches therefore expose the closed world of cultural ways into a wider nature, and provide some freedom in method and theory to explore outside the constraints of a strictly cultural life (see Costall 2004).

From structures to flow

If we are to adopt an archaeological approach grounded in direct action within environmental space, it is necessary to rethink some of the metaphors we use to describe the dynamics of social life. For a more inclusive description, construction metaphors seem too rigid and presumptuous to accommodate continuous activity. A better option relates to what unites all material phenomena – its energy. Like structure, energy is yet another metaphor, but it is dynamic; it accounts for the ability of a thing to do work. Matter is potential energy and motion is kinetic energy, which is released in the transformation of matter from one form into another. Life exists in these transformations, as they are the only way to "get anything done" (Smil 2017, 1).

Back in the gulch, the fragmented projectile point conjures up a description focused on action. The hunter is ready with a spear or atlatl, the act (searching, stalking, taking aim, hurling the shaft), is kinetic and skillful and, whatever the cultural context, the shaft finds its mark. As for the animal prey, walking and grazing outside culture along the stream, the effect is dire, dynamic and material. This creature, a member of its own community, as an indigenous hunter or biologist might say, had no idea about the vast conceptual world behind the projectile's strike.

A force that describes this energy transformation in ways suitable to action and perception is flow. This metaphor has a long history in descriptions of culture and society, but mostly without consideration as a system. Recently, however, Adrian Bejan has proposed *constructal flow* as a fundamental physical law (Bejan 2013, 25).

> The constructal law teaches us that nothing operates in isolation; every flow system is part of a bigger flow system, shaped by and in service to the world around it. The flow system we call a tree is also part of the larger flow system (that also includes rivers and weather patterns) for moving water from the ground to the air in order to achieve an equilibrium of moisture locally and globally. At the end of the day, the tree, like every other flow system, exists in order to facilitate nature's tendency to flow with configuration.

In the gulch, constructal flows are clearly at work – the running stream, continuously cutting deeper into the colluvium over time, the eroding, cross-cutting slopes, the swath of vegetation covering the stream margins, the winds, sunlight through the Ponderosa pines, the smells of pine needles, pitch, grass and damp earth and, by their traces, the actions of the hunter, artisan and archaeologist there. Flows of transformed energy course through all that we cannot see directly, from the electrical and chemical flows in brains and nervous systems to the endless shifting and erosion of supposedly inert rocks, stones and earth.

As metaphors of flow seem at home in the environment we experience, they may help realign ideas and evidence related to social behaviour. Turning back to the example of the spider, consider its central function in metaphor in a network, or a meshwork, as Ingold envisions (2010). The jumping spider would disagree about the web presenting its structure of possibilities, except that it detects in its environment an opportunity for a meal. The spider–web association is much more discrete and concrete than the network models with their intellectual implications – much closer to the concrete thing we see sketched out in the morning dew at the back of the garden. Depending on the species, spiders may use webs in defense (Blackledge, Coddington and Gillespie 2003), communication (Buskirk 1975), mating (Jackson 1986) and even as communal shelters, as some spiders build webs in the webs of other spiders (Jackson 1986). In this way it is more like the ground where we began, in the gulch, in which the flows of sediment and water downslope mediate the flows in the larger community of life that lives there.

This comparison is not about a difference in the horizons in the worlds of spiders, or about varieties of network designs; it implies the need to turn from representations as the ways and means of accounting for life's possibilities. When the spiders are grounded, the difference is between the one that builds and waits and the one that hunts. "Every individual 'flows' with the rest on the live landscape …" (Bejan 2013, 152).

The issue of scale

The difficulty with visualizing ground in the dynamics of cultural life is partly a matter of scale. Within the limits of history and direct experience, most of the sweeping geological changes in the earth happened so long ago and so slowly that they are invisible, and it might seem to have always been this way.

While the most recent glaciation that transformed so much of northern North America only began to retreat about 14,000 years ago, there is one place on earth, the Daintree Rainforest in Queensland, Australia, that has existed more or less unchanged for well over 100 million years. In comparison, humans are newcomers. Homo sapiens emerged within the past 500,000 years from its hominin cousins and ancestors, change from foraging and plant and animal management into agriculture took place within the past 10,000 years, and it has been almost 5500 years since the development of writing.

Thinking back to the gulch, by the time hunting and gathering people reoccupied the Great Plains of North America towards the end of the Pleistocene, the plant and animal life was already ancient. A common migrant that flies over the Black Hills on its annual route from northern Canada to Texas, the sandhill crane (Grus canadensis canadensis), is fossilized in deposits in Florida that are 2.5 million years old (Emslie 1992).

We have difficulties with scales of time because we fill our own with so much imaginative detail now that we have little time for life in the past then, and even less for the world outside humans, which carries on around us without the need for metaphor. Indeed, archaeologists speak about the lives of early humans as if 10,000 years of time then was the equivalent of a week of ordinary life in any community today. We should take the different scales of time and other material flows in the rest of the world into account, however, and not merely as side-trips on the ground or in the imagination. The reason is that physical systems, environments, materials and other creatures exist *within* our lives and are part of what shapes our endless preoccupations, even if they do not rise to awareness. To be able to gain insights into the effects of this silent world on cultural life, we must therefore face a dilemma usually left to philosophers and spiritual leaders, trying to understand culture from outside.

We know that there is a world out there that is larger than the extent of human life, one we experience but cannot express directly. If we remain open to its possibility, leave the comfort of narrative, and imagine ourselves as we see other forms of life, it may be much easier to identify commensurate behaviour over the gulf between past and present – to accept that people past and present exist on *shared ground.*

Of course, this challenges the dominion of cultural relativism and social self-identity and, indeed, we can only imagine the meanings of even the most mundane of objects and images today, let alone paintings from the Pleistocene, but there is little problem recognizing them in the first place. In an open environment, the solution here is to focus on the physical environment, and combine perceptual ecology, which tries to understand human perception, and sensory ecology, which explores perception in all living things.

Grounding culture

To find common ground in cultural life, imagine that culture is a kind of nature that sees itself. The problem then becomes, as ancient authorities imagined it in the Garden of Eden, a question of awareness. As we articulate experience within the horizons of language and our particular social inculcations, most of lived life – the energy and movement of our physical selves – is obscure. McLuhan several times used a not-so-cryptic aphorism to characterize this natural unawareness of material effects: "Fish don't know water exists till beached" (1970, 191). If we do not acknowledge these invisible forces, it seems impossible to understand the ways and means of ordinary existence.

The grounding of cultural life in a natural environment creates numerous issues in land-based fields such as archaeology and geography, especially with the most ubiquitous of representations used in studying the human use of space: landscape.

When the word first appears in print early in the 17th century, and long before the culture concept, landscape was a vista, a view of land from a distance or its representation in a painting and, in ordinary usage, the spatial setting of life. In other paintings, it was the decorative background, as Blount defines it in his dictionary in 1656: "All that which in a Picture is not of the body or argument thereof is Landskip, Parergon, or by-work" (Blount 1656). Since then, the landscape metaphor has become a way to accommodate cultural relativism and organize cultural activity within specific areas of land.

Early applications of landscape as a dynamic concept related to the cultural use of space began in geography, and the first approaches in archaeology were in this sense (e.g., Aston and Rowley 1974). In 1983 I proposed a regional map-based approach to rock art in the Canadian Shield in northern Canada. It would use existing archaeological site records, historical documents, oral traditions and ethnographic information to set up hypothetical distributions of significance that would inform archaeological surveys. Under the assumption that rock art was associated closely with religious belief, sacred landscapes would be "mapped and investigated like any other environment that influences human activity" (1983, 5).

Imagining landscape as an ideological veneer seemed useful for surveying the trackless, difficult terrain in the Canadian Shield, an area about twice the size of western Europe. As for interpretation, however, the idea of a simple correspondence of rock art and landscape ideology was complicated by an unanticipated discovery at a rock painting site – a small projectile point, visible in the mud on the lake bottom below a cluster of red ochre paintings. As the paintings were on a cliff that plunged into the water, it seemed reasonable that the point was placed there intentionally, either in a rock crevice or in the water below the images, but this provoked a troubling question. Which came first – an act of offering to a bare rock inhabited by spirits or to a painted rock that marked the presence of spirits? The study offered a lesson rather than a solution. Clearly, there was going to be no simple relation between a conceptualized landscape and the intentions of the people who lived in its bounds, and no simple match of rock art imagery and received traditions from other times, places and societies.

As the landscape concept developed, however, it gradually merged ideological and environmental space and went from being a guide for survey to an intellectual model that routinely characterizes human activity and guides interpretation. Ironically, it adopted Blount's term, landskip, but not his definition. Landscape now applied to what was important – expansive ideas about social meanings and functions in occupied space – while continuing to neglect the undescribed background, with its material flows and unyielding empirical evidence, still the by-work or parergon in cultural research.

During the rise of postmodernism in the late 20th century, it was perhaps inevitable that landscape would become theatre. In an edited volume published at the turn of the 21st century, landscape was an "… arena in which and through which memory, identity, social order and transformation are constructed, played out, re-invented, and changed" (Knapp and Ashmore 1999, 10). In effect, a landscape was a place that interpreters in the present constructed for people of the past to perform in. What is difficult to understand about this general approach to the earth is its preoccupation with ideas. Outside the conditions and horizons of landscape there is nothing, as if unacknowledged life and unappropriated things simply do not exist. With its existential authority, landscape is simply culture's environmental proxy.

This narrow focus should be a problem for a field archaeologist trying to deal with artifacts at a site. Overviews clearly form more general patterns, whether looking out from the heights or in the imagination. The problem is that models and representations also anticipate results, and they influence analysts to apply them to local situations, by smoothing out their variation from the norm or, worse, by treating their subjects as if they live in the model rather than in their surroundings.

Archaeology does need an approach that can account for multiscale patterns and variability, but not one with the implicit hierarchical bias of a great chain of being, in which cultural life as experienced is at the opposite end of a process of intellectual enquiry – in keeping with ancient concepts of human authority over nature. Bejan's constructal law states that "For a finite-size system to persist in time (to live), it must evolve in such a way that it provides easier access to the imposed currents that flow through it" (Bejan & Lorente 2010, 1). Thus, the flows of cultural life have an essential hierarchy, but such patterns have the freedom of the dendritic patterns of a stream or a tree, rather than structures imposed or rationalized from above.

In a discussion of prehistoric settlement patterns in the Mediterranean, Wandsnider (2004) offers a practical suggestion for regional archaeologists that fits with this more open concept of the experience of both original inhabitants and archaeologists: "rather than a record of settlements awaiting discovery and definition by the archaeologist, the archaeological record is better considered a record of places with different material histories, both cultural and natural" (Wandsnider 2004, 69).

Landscape and robotics

Some clarity in the conceptual muddle of environments, landscape and situations, and a possible way forwards, comes from a related field with similar concerns about the dynamics of environmental interaction – robotics.

The biggest design problem in late-20th century robotics was engineering human-like movements and functions in a spatial environment. In 1987, Rodney Brooks suggested that attempts to build machines to serve as human-like actors stalled because the focus of design was on the fit between the robot and a hypothesized human environment (Brooks 1987; 1991).

Creating an approximation of such an environment required the translation of the assumed properties of experience into concrete form with behaviours, obstacles and pathways known in advance, and this was never going to be possible. Ray Birdwhistell, the anthropologist who developed the study of cultural kinesics, showed how daunting the description was of even the simplest of tasks. He spent almost ten years studying and charting the kinetic movements in the 297 frames of a film depicting the lifting, lighting and smoking of a cigarette (1970, 220–50). So, what these engineers produced was more a map-like world for their robots that represented themselves, rather than the laboratory floor on which the robots would move. As Bourdieu (1977) observed in his *Theory of Practice*, a map is

> the analogy which occurs to an outsider who has to find his way around in a foreign landscape and who compensates for his lack of practical mastery, the prerogative of the native, by the use of a model of all possible routes.
>
> (1977, 2)

In his critique of existing design approaches, Brooks' conclusion provides the basis for an environmentally focused, situational approach to cultural life.

> When we examine very simple level intelligence we find that explicit representations and models of the world simply get in the way. It turns out to be better to use the world as its own model.
>
> (1991, 139)

Imagining the world from any robot's point of view, he engineered his to accomplish simple acts appropriate to *their* sensory systems and physical environment, such as moving away from obstacles they encounter. Such simple, direct responses to a physical world are endlessly repeated, despite the distractions of increasing cognitive complexity, in the sensory dynamics of organisms. By putting his robot in the world, actually and theoretically, rather than trying to make the robot the master of a world (cf. Barthes 1997), Brooks avoided the primal weakness of top-down models meant to approximate aspects of the human condition, which create worlds and look down on the people within them.

Models, plans and the theoretical tableaux of cultural landscapes are effective tools for creating order in dynamic processes, and for visualizing potential effects in the applied reality on the ground, just as a map eases navigation through unfamiliar territory. And just like a map, these abstract concepts are meant to help understand reality, not stand for it. A model that takes over its world has the effect of Erwin Goffman's "total institution" – a prison, hospital, asylum or anywhere the activity of a set of individuals is highly controlled (1961). As Goffman found, even in these most carefully controlled social environments, an underlife occurs in "standard places of vulnerability … the damp corners where secondary adjustments breed and start to infest the establishment" (1961, 268). The work-arounds of inmates in the underlife of institutions, the cracks in foundations, the exceptions to rules, the anomalous cases, all the creases that we smooth out with models, styles, statistics and explanations, are the local, direct and dynamic situations of daily life. Through these gaps in authority the world of direct experience shows through.

A situational approach

Brooks changed the direction of robotics by visualizing the situation that his robots would face in the environment available to them as perceivers and actors. In the world of flesh and blood, this does not provide us with a solution to the problem of representations, but it does inspire a change in attitude. It suggests that those who are preoccupied with mind and meaning as instrumental forces in cultural life open their eyes and look around at their surroundings. As we stand at archaeological sites or look at monuments or works of art, we are, like Brooks, at the best scale to understand cultural life as it was lived by others. At its simplest, as Dewey visualized, a situation is a pause in activity, in which one weighs circumstances and options, makes a choice and moves on. With no inherent scale and no necessary position in time or space, a situation is endlessly adaptable to material and social conditions, and so it has a significant value in a dynamic analysis. Situations imply local, discrete actions and outcomes – some of which leave material traces. The application of situational analysis in archaeology is therefore not in the service of ontology but rather a simpler form of curiosity, the quest to find out more about the society in which an object or feature emerged – picked up from the environment in which it was created.

To imagine the situations surrounding artifacts, it is necessary to change how we see them. The traditional idea of the object is a product independent of its maker, with descriptions of meaning and function retrieved from existing information remote from the source. A projectile point found on an eroded slope is all about culture and society, not the experience of its maker. On the contrary, the emphasis in a situational analysis is on the creative act in its immediate context. The culture is in the making, not the product.

As work takes place in a specific location and has a specific duration, the workflow is not a simple relation between intentions and results, governed by some higher concept – it emerges during the act, which combines randomness, habit and intent in a cascade of small decisions in response to contingencies of the moment. It is a *situation of production* (Molyneaux 1997b). And what comes out of a situation is an outcome, before it is a product. This is an *analogue object*, a direct, concrete record of the flows of material and social forces and information created through the actions of its maker.

The difference between outcome and product is illustrated well in the recording of audio signals (see Chicago School of Media Theory 2017; Krokowski 2017). In a traditional analogue recording, sound waves travel through a device that converts sound pressure directly into vibrations and transfers the vibrations onto a physical medium – a wax cylinder, a vinyl record or a magnetic tape. The source signal is direct and continuous, and so the record contains the signal it has received in the specific transmission, noise and all. In a digital recording, the sound waves are processed – intercepted, transformed into discrete units (bits) in binary code, sampled, filtered to remove noise and stored. The connection to the earth is replaced by an algorithm, which produces a resemblance of the original sound transmission. The noise outside its narrow intentions, intrusive life, is absent.

With its emphasis on origins, the situational approach therefore treats production as more than the simple prelude to the product and its latter-day cultural life. It gives renewed value to action and the primary work in the circumstances of creation of artifacts. In a crafter's workshop, for example, where commissions, desires, sketches or plans may guide the work and deadlines may loom, what brings a picture, pot or projectile point into being is the working, the play of movement and ideas, the shaping of it. Writing about a basket weaver, Tim Ingold describes the "force field, in which the weaver is caught up in a reciprocal and quite muscular dialogue with the material" (Ingold 2000, 342). Making any kind of thing is a way of moving through a flow environment that leaves concrete evidence of its actions. In the moments of vision and movement, the external world shows through, leaves its marks and then quickly retreats as the paint dries, the pot is fired, the last flake falls and the newly transformed matter joins in its social flow.

A situational approach also requires a turn from the search for meaning. In place of the idea that an individual maker is a social proxy who unwittingly expresses some cultural or social dynamic conceived in remote discourse, a situational analysis considers the social act at the maker's scale, as the discrete set of actions over the duration of production. And this exposes the individual anarchy of creation. Consider the American painter Jasper Johns' description of his working process (Welish 1996).

> Whatever form the work takes—objects or non-objects or one type of mark or division or others—on some level it all ends up as thought. There are weird shifts among object, subject, perceiver, and what's perceived. And thought can seem of greater concern than the presence of the object. Artists play with what comes into their minds and with the materials at hand, and there is naturally feedback into their minds and into the painting process. So at any moment something may be dropped or brought in.

If this is what a situational analysis entails for a typical artist or crafter in the process of creation, it may seem impossibly subjective. However, in the moments when Johns or any other artist or crafter draws back from the unreflective parts of the task, they see with spectators' eyes. It is this convergence of kinetic action, perception, vision, skill and social experience that brings the work to its conclusion – and gives it a place in art history and archaeology.

To explore this rapprochement with nature, we begin with a brief look at a monument built in an old battlefield and then move on to two great natural monuments that have galvanized attention and provoked ritual and belief over thousands of years. The past world an archaeologist encounters, no matter how it might be represented, is a cumulation of individual perceptions and ideas resolved in concrete acts: speech and the objects, materials and impacts of work. The best reason for taking a situational perspective may therefore be that it is appropriate to its data – the direct, immediate, tangible experiences of ordinary life.

References

Aston, M., and T. Rowley. 1974. *Landscape Archaeology: An Introduction to Fieldwork Techniques on Post-Roman Landscapes.* Newton Abbot: David and Charles.

Barrett, Louise. 2011. *Beyond the Brain: How Body and Environment Shape Animal and Human Minds.* Princeton: Princeton University Press.

Barthes, Roland. 1997. *The Eiffel Tower, and Other Mythologies.* Berkeley: University of California Press.

Bejan, Adrian. 2013. Culture and the Constructal-Law evolution of the Human and Machine Species. Comment on "An evolutionary framework for cultural change: Selectionism versus communal exchange" by L. Gabora. *Physics of Life Reviews* 10:151–153.

Bejan, A., and S. Lorente. 2010. The Constructal Law of Design and Evolution in Nature. *Philosophical Transactions of the Royal Society of London.* Series B, Biological Sciences, 365 (1545), 1335–1347. https://doi.org/10.1098/rstb.2009.0302.

Birdwhistell, Ray. 1970. *Kinesics and Context: Essays on Body Motion Communication.* Philadelphia: University of Pennsylvania Press.

Blackledge, Todd A. , Jonathan A. Coddington and Rosemary G. Gillespie. 2003. Are Three-Dimensional Spider Webs Defensive Adaptations? *Ecology Letters* 6:13–18.
Blount, Thomas. 1656. *Glossographia, or, A Dictionary Interpreting all such Hard Words of Whatsoever Language now Used in our refined English Tongue with Etymologies, Definitions and Historical Observations on the Same: also the terms of Divinity, Law, Physick, Mathematicks and other Arts and Sciences Explicated.* London: Tho. Newcomb.
Bourdieu, Pierre. 1977. *Outline of a Theory of Practice.* Cambridge: Cambridge University Press.
Brooks, Rodney A. 1987. Planning is Just a Way of Avoiding Figuring Out What to Do Next. *Working Paper* 303. Cambridge: Artificial Intelligence Laboratory, Massachusetts Institute of Technology.
Brooks, Rodney A. 1991. Intelligence Without Representations. *Artificial Intelligence* 47:139–159.
Buskirk, Ruth E. 1975. Aggressive display and orb defence in a colonial spider, Metabus gravidus. *Animal Behaviour* 23:560–567.
Chicago School of Media Theory. 2017. *Analog/Digital.* https://lucian.uchicago.edu/blogs/mediatheory/keywords/analogdigital/.
Copernicus, Nicholas. 1543. *De revolutionibus orbium coelestium* (On the Revolutions of the Heavenly Spheres). Nuremberg: Johannes Petreius.
Costall, Alan. 1984. Are Theories of Perception Necessary? A Review of Gibson's The Ecological Approach to Visual Perception. *Journal of the Experimental Analysis of Behavior* 41:109–115.
Costall, Alan. 1985. How Meaning Covers the Traces. In *Visual Order, the Nature and Development of Pictorial Representation*, edited by N.H. Freeman and M.V. Cox, 17–30. Cambridge: University of Cambridge Press.
Costall, Alan. 1986. The "Psychologist's Fallacy" in Ecological Realism. *Teorie & Modell* III:37–46.
Costall, Alan. 2004. From Darwin to Watson (and Cognitivism) and Back Again: The Principle of Animal-Environment Mutuality. *Behavior and Philosophy* 32:179–195.
Costall, Alan. 2007. How Cognitive Psychology Highjacked Thinking. *Anthropological Psychology* 18:21–23.
Cottingham, John, Robert Stoothoff and Dugald Murdoch, trans. 1985. *The Philosophical Writings of Descartes*, Volume 1. Cambridge: Cambridge University Press.
Cresswell, Richard, trans. 1887. *Aristotle's History of Animals. In Ten Books.* London: George Bell & Sons.
Darwin, Charles. 1859. *On the Origin of Species by Means of Natural Selection, Or the Preservation of Favoured Races in the Struggle for Life.* London: John Murray.
Darwin, Charles. 1871. *The Descent of Man.* London: John Murray.
Day, Joy, ed. 2013. *Making Senses of the Past: Toward a Sensory Archaeology.* Carbondale: Southern Illinois University Press.
Debord, Guy. 1995. *The Society of the Spectacle.* Translated by Donald Nicholson-Smith. New York: Zone Books.
Defoe, Daniel. 1726. *Mere Nature Delineated: or, a Body without a Soul. Being Observations upon the Young Forester Lately Brought to Town from Germany. With Suitable Applications. Also, a Brief Dissertation upon the Usefulness and Necessity of Fools, whether Political or Natural.* London: T. Warner, at the Black Boy, in Pater-Noster-Row.

Dimmitt, Cornelia, and J.A.B. Van Buitenen. 1978. *Classical Hindu Mythology. A Reader in the Sanskrit*. Philadelphia: Temple University Press.

Dusenbery, David. 1992. *Sensory Ecology: How Organisms Acquire and Respond to Information*. New York: W.H. Freeman.

Emslie, S.D. 1992. Two New Late Blancan Avifaunas from Florida and the Extinction of Wetland Birds in the Plio-Pleistocene. *National History Museum of Los Angeles County. Contributions in Science* 36:249–269.

Ghosh, Sanjib Kumar. 2015. Human cadaveric dissection: a historical account from ancient Greece to the modern era. *Anatomy and Cell Biology* 48:153–169.

Gibson, James J. 1966. *The Senses Considered as Perceptual Systems*. Boston: Houghton Mifflin.

Gibson, James J. 1979. *The Ecological Approach to Visual Perception*. Boston: Houghton Mifflin.

Goffman, Erving. 1961. *Asylums: Essays on the Social Situation of Mental Patients and Other Inmates*. Garden City: Anchor Books.

Harland, Duane P., and Robert R. Jackson. 2000. "Eight-legged Cats" and How They See – a Review of Recent Research on Jumping Spiders (Araneae: Salticidae). *Cimbebasia* 16:231–240.

Ingold, Tim. 2000. *The Perception of the Environment: Essays on Livelihood, Dwelling and Skill*. London: Routledge.

Ingold, Tim. 2010. *Bringing Things to Life: Creative Entanglements in a World of Materials*. ESRC National Centre for Research Methods, Working Paper #15. Manchester: University of Manchester.

Jackson, Robert R. 1986. Communal Jumping Spiders (Araneae: Salticidae) from Kenya: Interspecific Nest Complexes, Cohabitation with Web-building Spiders, and Intraspecific Interactions. *New Zealand Journal of Zoology* 13:13–26.

Jackson, R.R., and D.P. Harding. 1982. The biology of Holoplatys sp. indet., a New Zealand Jumping Spider (Araneae: Salticidae): Intraspecific Interactions. *New Zealand Journal of Zoology* 9:487–510.

Knapp, A. Bernard, and Wendy Ashmore. 1999. Archaeologies of Landscape: Contemporary Perspectives. In *Landscape. Contemporary Perspectives*, edited by Wendy Ashmore and A. Bernard Knapp, 1–30. Oxford: Blackwell.

Krokowski, Damon. 2017. *The New Analog. Listening and Reconnecting in a Digital World*. New York: The New Press.

Lambert, Wilfred G., and Alan R. Millard. 1969. *Atra-hasis: The Babylonian Story of the Flood*. Oxford: Oxford University Press.

Latour, Bruno. 1993. *We Have Never Been Modern*. Translated by Catherine Porter. Cambridge: Harvard University Press.

Layton, Robert, and Peter J. Ucko. 1999. *The Archaeology and Anthropology of Landscape: Shaping Your Landscape*. London: Routledge.

Linnaei, Caroli. 1758. *Systema Naturae per Regna Tria Naturae, Secundum Classes, Ordines, Genera, Species, cum Characteribus, Differentiis, Synonymis, Locis. Tomus I. Editio Decima, Reformata*. Holmiae [= Stockholm]: Impensis Direct. Laurentii Salvii.

Locke, John. [1690] 1721. *An Essay Concerning Human Understanding*. 8th Edition. Printed for A. Churchill and A. Manship and sold by W. Taylor at the Ship and Black Swan.

Lovejoy, A.O. 1936. *The Great Chain of Being*. Cambridge: Harvard University Press.

Lyell, Charles. 1830. *Principles of Geology, being an attempt to explain the former changes of the Earth's surface, by reference to causes now in operation*. London: John Murray.

McLuhan, Marshall. 1962. *The Gutenberg Galaxy: The Making of Typographic Man*. Toronto: University of Toronto Press.

McLuhan, Marshall. 1964. *Understanding Media: the Extensions of Man*. New York: McGraw-Hill.

McLuhan, Marshall. 1970. *Culture Is Our Business*. New York: Balantine Books.

Mead, George H. 1934a. *Mind, Self and Society. From the Standpoint of a Social Behaviorist*. Edited by Charles W. Morris. Chicago: University of Chicago.

Mead, George H. 1934b. *The Philosophy of the Act*, edited by Charles W. Morris. Chicago: University of Chicago Press.

Menda, Gil, Paul S. Shamble, Eyal I. Nitzany, James R. Golden and Ronald R. Hoy. 2014. Visual Perception in the Brain of a Jumping Spider. *Current Biology* 24:2580–2585.

Merleau-Ponty, Maurice. 2008. La Nature ou le Monde du Silence. In *Maurice Merleau-Ponty*, edited by Emmanuel de Saint Aubert, 44–53. Paris: Hermann Philosophie.

Molyneaux, Brian. 1983. *The Study of Prehistoric Sacred Places: Evidence from Lower Manitou Lake*. Archaeology Paper 2. Toronto: Royal Ontario Museum.

Molyneaux, Brian L., ed. 1997a. *The Cultural Life of Images: Visual Representation in Archaeology*. London: Routledge.

Molyneaux, Brian L. 1997b. Introduction: the Cultural Life of Images. In *The Cultural Life of Images: Visual Representation in Archaeology*, edited by Brian L. Molyneaux, 1–10. London: Routledge.

O'Keefe, J., and J. Dostrovsky. 1971. The Hippocampus as a Spatial Map. Preliminary Evidence from Unit Activity in the Freely-moving rat. *Brain Research* 34:171–175.

O'Keefe, John, and Lynn Nadel. 1978. *The Hippocampus as a Cognitive Map*. Oxford: Oxford University Press.

Richman, David B., and Robert J. Jackson. 1992. A Review of the Ethology of Jumping Spiders (Araneae, Salticidae) from Queensland. *Bulletin of the British Arachnological Society*1992(2):33–37.

Sims, M.E. 2007. *Comparison of Black Bear Paws to Human Hands and Feet*. Identification Guides for Wildlife Law Enforcement No. 11. Ashland: USFWS, National Fish and Wildlife Forensics Laboratory.

Singer, Charles, trans. 1956. *Galen on Anatomical Procedures*. London: Oxford University Press.

Smil, Vaclav. 2017. *Energy and Civilization. A History*. Cambridge: MIT Press.

Tarsitano, Michael S., and Robert R. Jackson. 1997. Araneophagic Jumping Spiders Discriminate between Detour Routes that Do and Do Not Lead to Prey. *Animal Behaviour* 53:257–266.

Tsang, A.M., and N.E. Klepeis. 1996. *Descriptive Statistics Tables from a Detailed Analysis of the National Human Activity Pattern Survey (NHAPS) Data*. Washington: U.S. Environmental Protection Agency.

Tylor, Edward. 1871. *Primitive Culture: Researches into the Development of Mythology, Philosophy, Religion, Art, and Custom*. London: John Murray.

Tyson, Edward. 1699. *Orang-Outang, Sive Homo Sylvestris: or the Anatomy of a Pygmie Compared with that of a Monkey, an Ape and a Man. To which is added, a Philological Essay Concerning the Pygmies, the Cynocephali, the Satyrs, and Sphinges of the Ancients. Wherein it will appear that they are all either apes or monkeys, and not men, as formerly pretended*. London: Printed for Thomas Bennet at the Half-Moon in St. Paul's Church-yard; and Daniel Brown at the Black Swan and

Bible without Temple-Bar and are to be had of Mr. Hunt at the Repository in Gresham-Colledge.

Venturi, Robert. 1977. *Complexity and Contradiction in Architecture*. Second Edition. New York: Museum of Modern Art.

Venturi, Robert, Denise Scott Brown and Steven Izenour. 1972. *Learning from Las Vegas*. Cambridge: MIT Press.

Vesalius, Andreas. 1543. *De Humani Corporis Fabrica Libri Septe*. Basel: Johannes Oporinus.

Von Uexküll, Jacob. [1934] 1957. A Stroll Through the Worlds of Animals and Men. In *Instinctive Behavior, The Development of a Modern Concept*, translated and edited by Claire H.Schiller, 5–81. New York: International Universities Press.

Wandsnider, LuAnn. 2004. Artifact, Landscape, and Temporality in Eastern Mediterranean Archaeological Landscape Studies. In *Mediterranean Archaeological Landscapes: Current Issues*, edited by Effie F. Athanassopoulos and LuAnn Wandsnider. Philadelphia: University of Pennsylvania Museum of Archaeology and Anthropology.

Warren, William W. [1885] 1984. *History of the Ojibway People*. St. Paul: Minnesota Historical Society Press.

Welish, Marjorie. 1996. Jasper Johns. *Bomb* 57. https://bombmagazine.org/articles/jasper-johns/.

White, Leslie. 1949. *The Science of Culture. A Study of Man and Civilization*. New York: Grove Press.

Zitkala-Sa. 1985. *Old Indian Legends. Retold By Zitkala-Sa*. Lincoln: University of Nebraska Press.

3 The lure of monuments

Space, place and situation

When Yi-Fu Tuan asks "What is the nature of the objective space over which human beings have variously projected their illusions?", he accepts as facts that "The space that we perceive and construct, the space that provides cues for our behaviour, varies with the individual and cultural group" (Tuan 1979, 389). This comfortable accommodation reflects the traditional use of cultural models and representations in the understanding of experience. Yet the body in its physical space knows nothing of this. What is missing is the redolent nature outside meaningful horizons, the environmental interactions that implicate humans and all other creatures in the flows of energy and matter that shape physical and social worlds.

Awareness of omnipresent forces that exist without a special nod to humans is an ancient insight, evident in a wide range of philosophies. It is clear in the first lesson of the Tao Te Ching (Lin 2006, 3):

> The Tao that can be spoken is not the eternal Tao
> The name that can be named is not the eternal name
> The nameless is the origin of Heaven and Earth
> The named is the mother of myriad things.

In the Great Plains of South Dakota, a Presbyterian missionary seeking converts among the Sioux tribes living along the Missouri River struggles to explain the concept of *wakan.*

> All life is Wakan. So also is everything which exhibits power, whether in action, as the winds and drifting clouds, or in positive endurance, as the bowlder by the way side.
>
> (Riggs 1869, 56–57)

This pragmatic directness, to acknowledge that which exists in bare awareness, is implied in Dewey's observation that "It is not experience which is experienced, but nature" (1929, 419).

DOI: 10.4324/9781351053105-3

To visualize this concept, consider the dynamics of proprioception. This is the system of sensory receptors in muscles, joints, tendons and other parts of the body that enables controlled movement in space. It is difficult to visualize its reckoning of location, but this peculiar ability enables a sense of position and direction, balance and sensations of force, mass and weight, and its results are manifest. Animals could not move coherently without it; even jumping spiders retain memory of their position in space, and appear to use proprioceptive information to navigate back to a nest in darkness (Taylor 1995). It is thus an example of a palpable, natural sense in all active creatures of something present, but without apparent substance, that either affords or constrains intentional activity. When you reach out for an object without looking at it, the dim awareness that completes the grasp is a sense of these invisible forces at work.

The idea that humans are immersed *within* physical forces has long been expressed in literary battles with the elements. While scholars struggle to make sense of the world with empirical methods, models and other analytical devices, fiction has the licence to resolve uncertainty through the imagination. Edgar Allan Poe explores this problem through the tale of a man and his brother, sailing in a schooner-rigged smack off the coast of Norway, who are driven by a hurricane into a great whirlpool (*A Descent into the Maelstrom*, 1841). While the boat slips down the whirling sides of the vortex, ever closer to the abyss, and the brother hangs on in terror to an iron ring on the deck, the protagonist looks carefully at its fearsome nature and finds a way to take advantage of its power. He lashes himself to an empty barrel, abandons the ship and goes with the flow, until the vortex subsides and he is cast out like jetsam. Only he survives.

This suggests that we are immersed in natural forces we have some capacity to master, and that what counts is how we adapt. Karl Marx (1906, 197–98), found this integration in labour, an act in which mankind uses nature as a tool on itself.

> He opposes himself to Nature as one of her own forces, setting in motion arms and legs, head and hands, the natural forces of his body, in order to appropriate Nature's productions in a form adapted to his own wants.

Lewis Mumford emphasized the opposite aspect, anticipating the loss of authority over nature, but emphasizing how tools also shape individuals and societies (1934, 321).

> The tools and utensils used during the greater part of man's history were, in the main, extensions of his own organism; they did not seem to have—what is more important they did not seem to have—an independent existence. But though they were an intimate part of the worker, they reacted upon his capacities, sharpening his eyes, refining his skill, teaching him to respect the nature of the material with which he was dealing. The tool brought man into closer harmony with his environment, not

> merely because it enabled him to re-shape it, but because it made him recognize the limits of his capacities. In dream, he was all powerful: in reality he had to recognize the weight of stone and cut stones no bigger than he could transport.

In these examples, energy reveals itself as capable of being controlled and transformed through human ingenuity and, simultaneously, mediating perception and action. Together, these descriptions express a practical sense of integration that in their directness and simplicity may provide some needed regrounding in studies of culture and society.

The issue is our growing distance from physical nature and direct interaction. On the far side of the industrial revolution, a little more than 200 years after Poe imagined the interaction of humans and nature, machines have taken over most of the work. People once directly engaged in production are now being served, and social interaction is now more digital engagement than flesh and blood. In a disturbingly practical way, technology has replaced God and culture in authority over social life.

This distancing seems almost ordinary now, and superior to old-fashioned kinds of engagement. Online life confers a wonderful mastery over information and an ability to marshal people and ideas that easily reaches across the globe, and this instrumentality lends a sense of reality to online experience. As with prayer or gossip, digital speech and remote social interaction can shape events and sometimes have real world effects for the better, and this validates their existence.

At the same time, the ideological space is so populated by avatars, ideas and representations that old connections with the earth might as well be broken – even the way we relate to other people in social bonds, a dynamic that is more ancient than the human species, is increasingly remote. The effects extend throughout the arts, humanities and social sciences – and here, in archaeology. While the stereotypical archaeologist finds and studies artifacts, much archaeological research today is consumed in discourse about a domain far outside the modest physical evidence and the people who created it.

Yet, if physical life is an interaction of people and their surroundings, reality must also take shape within this sensed, but not necessarily seen, environment. Indeed, as existential threats such as global warming and the Covid-19 pandemic show, freedom from nature is all in our heads. Research into the ideological aspects of culture and society cannot therefore ignore the physical world and simply hand it to the sciences, pretending it has no part in cultural life. It seems increasingly urgent that we turn our attention once again from social self-possession to this inconvenient nature, the indifferent and sometimes dangerous forces that people have faced and implicitly understood since time immemorial. After all, ancient thinkers knew that speaking about the forces of nature was a problem – because its forces defy description – but they did not neglect the world outside their own speech. They recognized that human beings were part of the surroundings from which they emerged.

Imagine that an archaeologist, a Pleistocene hominin, a dog and a crow, traveling together, suddenly came to the ledge above our gulch in the Black Hills, the one where the Paleoindian hunter waited for game to pass by below. All had to stop but the crow, who paused and then flew on. Every creature has specific perceptual horizons and faces its own form of abyss, but in an environment it shares with others, they are united – all without philosophy or instruction.

With its interests in artifacts and materiality, archaeology may be able to help renew interest in this world of direct experience. Not only artifacts, materials and traces of action survive time and neglect in the places we find them. So does the physical environment that exists within our capacity as human beings. If we use this persistent world as a model, research about the past is not limited to rationalized social relations, projected in representations and applied to the artifacts and traces below. As our story of the gulch and the mountain in the Black Hills suggests, places may retain the affordances that attracted people there in the first place.

Because this human–environment relationship is fundamental to cultural life, the case studies ahead will explore the implications of our discussion so far in physical settings that have an explicitly ideological aspect, ranging from sacred monuments to landscape paintings. As the goal is to sketch out a field of possibilities for future research, the discussion ranges within several disciplines, and stays on the surface of many complex issues. The lure of situational analysis for an archaeologist is that while meaning is ephemeral, the shapes that meaning has taken are preserved in the objects, features and their relations that trace past cultural life. If production is a situation involving physical and social urges and imperatives, and the artifacts and relations are analogues of these actions, there may be more to find about past cultural life in the outlands, fields and settlements than one might think.

Butte du Leon, Lion's Mound

On June 16, 1815, south of the hamlet of Quatre Bras in the gently rolling Belgian countryside, a force of allied Dutch and British troops fought a French army intent on opening the road to Brussels, about 60 km to the northeast. The casualties in this skirmish, dead and injured, totalled almost 10,000. Two days later, almost half-way along this road near the village of Waterloo, the armies of Wellington and Napoleon fought the climactic battle that ended Napoleon's ambitions for France in Europe. In less than 12 hours, almost 50,000 soldiers were dead or injured.

Both forces intruded here and battled for purposes remote to these fields and, when it was over, they quickly removed, leaving the aftermath to local inhabitants and the resonance to work its way across the world. The bodies were buried or disposed of, possessions, ordnance and equipment scavenged and, eventually, farmers again ploughed, laid crops and picked up life as it was the day before the battle. The rest was left for scavengers and, eventually, archaeologists.

Outside the fields, however, the imaginary battle had just begun. In this "visionary mist" (Hugo 1887, 76), through which we all see the battle, plans and strategies, movements and events continue to be pondered and debated, and more than 200 years later, the discourse overwhelms the raw evidence of what took place in those few hours in 1815.

There is also something about harrowing events that demands a more tangible response. In the aftermath of battles, the indifference of nature seems intolerable, and so in the longer term there is a need to resist the slow erosion of memory in cultural life. For private, devastating grief, there are mementos, pictures and sometimes graves; but for the public, there must be monuments.

Between 1823 and 1826, King William I of the Netherlands had a large, conical earthen mound constructed in the battlefield – it is said, in the place where his son, the Prince of Orange, was wounded by a musket ball in the shoulder (Silberman 2007, 56). Workers scraped up the battlefield around the site for the dirt, destroying the military roads and levelling the rise where Wellington stood in the battle. Soldiers were sacrificed a second time, as many bodies were disturbed, as an eyewitness, Reverend William Falconer, who visited the site in 1825, reported (Uffindell & Corum 1966, 34).

> A great deal of earth is of course collected from the parts close around and to serve their purpose they have lately levelled the rising ground where the Duke stood during the contest. It was about there I picked up a human rib just disturbed from its resting place.

The vainglorious result was topped by a huge cast-iron lion with its paw resting on a globe, facing east towards France – at the very least, signifying the triumph of the Netherlands and Great Britain over France.

Major Richard Blackiston visited the site in 1833, and he lamented the destruction (Blackiston 1835, 19–20):

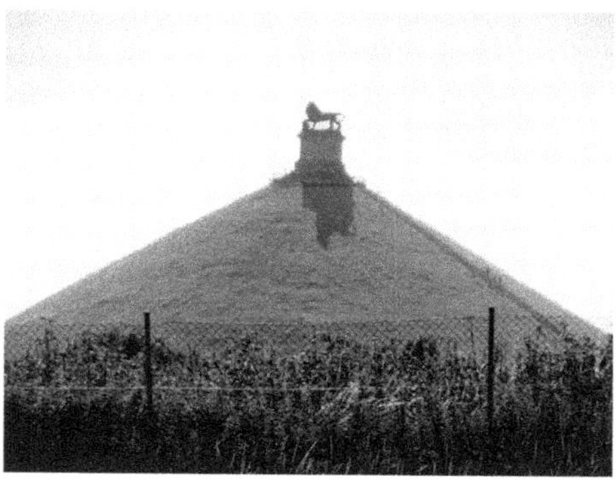

Figure 3.1 The Lion's Mound

> This mass of earth, surmounted by a lion, has been formed by scraping off about ten or twelve feet of the soil from nearly half a mile's length of the British position, which it has completely defaced; for the inequality of the ground was so little, that a few feet taken off from the surface has made a material difference in its appearance. A smaller pyramid of stone would, in my opinion, have had a better effect. If it was intended as a favourable spot from which to view the field, the object has failed; for when standing on such an eminence the inequalities of the ground are scarcely perceptible, and no defined position can be made out.

Hugo commented in 1861 on the monumental irony (Hugo 1887, 321).

> By taking from this mournful field the wherewithal to make a monument to it, its real relief has been taken away, and history, disconcerted, no longer finds her bearings there. It has been disfigured for the sake of glorifying it.

The king honoured his son's heroism and the glory of the victors with the Lion's Mound to give the battle a more permanent salience – simple, direct and concrete features, authentic or not, that visitors could see. In the years following, a scatter of small monuments across the whole theatre of battle in the region has created what Silberman calls a "landscape of memory" (2007, 56) and, since the 200th anniversary, this landscape is growing as buildings of the time are restored and more museums and displays open – even though there is neither landscape nor memory among these fields and settlements in the 21st century.

Still, the large and aggressive Lion's Mound conducts an unending battle with the lambent fields around it to ensure that some impression of the old is not entirely lost – and, if it is, the monument will continue to feed the imagination of those who see it in novel ways. This is so because, while a monument is significant because of the ideas and emotions it evokes, it also provokes these ideas and emotions because of the way its form and mass draw attention.

The idea of a physical Lion's Mound that knows not lions or battles, but provokes the people who do, describes a type of energy flow that is repeated over the topographies of the world, anywhere with a feature that draws attention for some aspect of its physical form or resources. The two monuments that follow are both natural features that show these forces in action among diverse cultures, over a vast territory and thousands of years.

Samanakuta, Sri Pada, or Adam's Peak

Visit of the Buddha

On the island of Sri Lanka, at the southwestern edge of the central highlands, a distinctive pinnacle rises high above the thick rainforests. What makes it so striking is its peak: a steep pyramid of bare rock converging in a small, rounded summit.

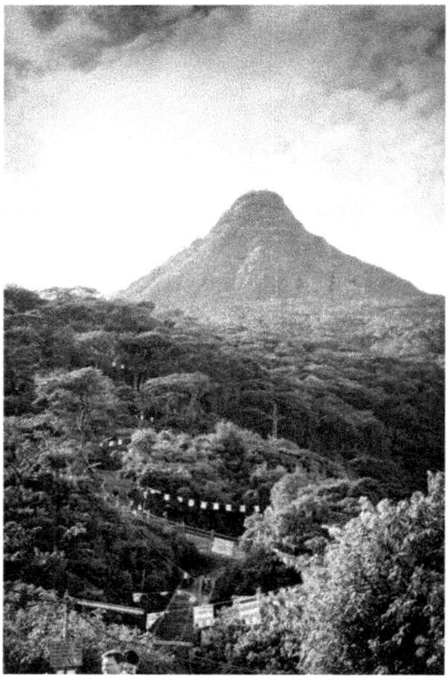

Figure 3.2 Samanakuta

The 4th century CE Theravada Buddhist chronicles of the Island, the Dipavamsa (Oldenberg 1879) and the Mahavamsa (Geiger 1912) name this pinnacle Samanakuta (the abode of the deva Mahasumana). The Mahavamsa begins with legendary accounts of the travels of Gautama Buddha to the Island, and it relates that when the Buddha had completed his task he left "the traces of his footsteps plain to sight on Sumanakuta" (I:78 in Geiger 1912, 8).

The story of this footprint must have been widely shared. Fu-Hsien, a Chinese Buddhist who visited the island ca. 410 CE, heard a more detailed account: "When Buddha came to this country wishing to transform the wicked nagas, by his supernatural power he planted one foot at the north of the royal city [Anuradhapura], and the other on the top of a mountain" (Legge 1886, 102) – a naga was a divine creature in Hindu and Buddhist belief, with snake and human features and capabilities. North of the Island, in Tamil Nadu, the epic Tamil poem, Manimekalai (ca. 6th century CE), refers to this remarkable event as footprints (Cattanar and Pandian 1989).

The travels of the Buddha ground beliefs in ways similar to the supernatural beings in many indigenous cultures who shaped the world in their exploits. Ordinary people may follow his paths, see his footsteps and draw in the surrounding sights and smells of the places he visited.

Outside the Mahavamsa, there are few references to the Peak for several centuries (see Paranavitana 1958), but this is not because it had been forgotten. With the Buddha's footprint on its summit, there was nothing more to say. It became a holy place and a pilgrimage destination. Near the end of the

11th century, however, political necessity seems to have brought the Peak back to public awareness. Beginning in 993 CE, the Chola Dynasty on the adjacent mainland expanded its rule to the north of the Island (Sastri 1955, 172), and for slightly more than a decade, between 1029 and 1041 CE (1955, 200), the southwestern part of the Island, known in the Mahavamsa as Rohana, conducted "an incessant warfare for the restoration of Sinhalese independence" (1955, 253). Resistance continued until 1071, when the Sinhalese, under Prince Kitti, a member of the deposed Sinhala royal family, succeeded in pushing the Cholas back to the mainland. After this victory, he took the name Vijayabahu I (1059–1114 CE), ruler of the whole Island, and in 1111 CE had the account of his victory inscribed on rocks at Ambagamuva, which is located in the highlands north of the Peak.

This was also the end of the natural appearance of the summit. The Culavamsa (the continuation of the Mahavamsa chronicle) records that Vijayabahu climbed to the summit of Samanakuta, venerated the footprint and spent the night there, and subsequently had rest-stations and alms houses built for pilgrims, enclosed the summit with a terrace and provided a lower-level structure from which low-caste people could worship the relic (Geiger 1929, 173–74).

The Peak once more appeared in historical accounts during resistance associated with the harsh Portuguese occupation in the 1580s. Raja Sinha I, the ruler of the highland kingdom of Kandy, responded to a plot against him involving members of the Buddhist royal court and priesthood by removing them from control over the Peak in favour of Hindu priests (Pieris 1920, 94–95). This state of affairs persisted until 1753, during the Dutch coastal occupation, when Kirti Sri Rajasinha brought Buddhist priests from Siam to restore Buddhist institutions and temples (Holt 1996).

After this restoration, the Peak seems to have settled into a state of peaceful worship, with a Buddhist shrine enclosing a footprint that Hindus and Muslims see through their own origin stories and beliefs.

Every year, thousands of pilgrims make the arduous trek along one of two major routes to the summit, beginning after the first moon in January (Durutha poya day) and Vesak, the full moon in May, which celebrates Gautama Buddha's birth, enlightenment and death. A local economy serves the pilgrims, offering spiritual support, food, water and accommodation on the lower flanks of the Peak and along the trails.

The Peak against the sky

The Peak has its own nature outside the footprint, the way it soars above the horizon of mountains at the southwestern edge of the highlands when viewed from the sea. This aspect may have inspired the Muslim belief that the hollow in its summit was Adam's footprint, left when he came down from Paradise to the earth.

By the time of the chronicles, there was a thriving sea trade along the coastal arch of the southern ocean, from east Africa to southeast Asia, and

Sri Lanka was central to this trade. About 457 AD, Cosmas Indicopleustes, a merchant who had become a Nestorian monk, wrote about this trade in a succinct and insightful way in his treatise, *Christian Topography* (McCrindle 1897, 365–66).

> The island being, as it is, in a central position, is much frequented by ships from all parts of India and from Persia and Ethiopia, and it likewise sends out many of its own. And from the 366 remotest countries, I mean Tzinista [China] and other trading places, it receives silk, aloes, cloves, sandalwoods and other products, and these again are passed on to marts on this side, such as Male, where pepper grows, and to Calliana which exports copper and sesame-logs, and cloth for making dresses, for it also is a great place of business. And to Sindu also where musk and castor is procured and androstachys, and to Persia and the Homerite country, and to Adule. And the island receives imports from all these marts which we have mentioned and passes them on to the remoter ports, while, at the same time, exporting its own produce in both directions.

Arab merchants dominated this early sea trade (Hirth 1896). They lived in a most favourable location, at the axis between Mediterranean Europe, Africa and west Asia, but most importantly, they invented a sea-going dhow that could take a more direct route across the open waters of the Arabian Sea and the Bay of Bengal, avoiding the coasts. The crossing was dangerous, but they were free of the threat of pirates, and so it was a risk worth taking, and it enabled them to serve the ambitions and demands of regional kingdoms and empires, east and west. By the rise of Islam, Arab traders had already established import–export firms in a number of Chinese cities (Lipman 1997, 27).

Because sailors had to navigate the open seas without the aid of compasses, sighting landmarks was essential. After passing the island chain of the Maldives, they had to spot the island of Sri Lanka, almost 150 leagues to the east. If they passed too far south, they were lost in the emptiness of the Indian Ocean. In such a voyage, the sighting of the Peak was crucial. In the words of Ibn Batuta, sailing east from the Maldive Islands in 1314, this mountain appeared "like a pillar of smoke, when we were at a distance of nine days from it" (Lee 1829, 183).

It was probably after the onset of Islam that merchants learned of the footprint on the summit. There was no question of accepting it as the Buddha's; they sought answers in their own holy texts. The Bagdhadi scholar Hisham Ibn-Al-Kalbi (737–819 CE) wrote on historical authority that the place where Adam came down from the heavens was in India (Faris 1952, 43).

> When Adam died the children of Seth (Shith), the son of Adam, buried him in a cave in the whereon Adam alighted (when he was sent) to the land of India. The name of the mountain is Nawdh, and it is the most fertile mountain in all the world.

Perhaps to Arab sailors of the 8th century following the trade routes, Nawdh was this mountain. What is clear is that the Peak begins to appear in Arab accounts of the sea trade beginning in the 9th century with an account generally attributed to Sulaiman al-Tajir, a Persian merchant, in ca. 850 CE (Mackintosh-Smith 2017, 25).

> In the territory of Sarandib is a mountain called al-Rahun. It is on this that Adam descended, eternal peace be upon him, and his footprint is on the bare rock of the summit of this mountain, impressed in the stone. There is only one footprint at the summit of this mountain, but it is said that Adam, eternal peace be upon him, took another step into the sea. It is said, too, that the footprint on the summit of this mountain is about seventy cubits long. Around this mountain lies the area where gems are mined—rubies, yellow sapphires, and blue sapphires.

The Baghdadi scholar Al-Masudi (ca. 896–956) also writes of this episode, weaving more of the Arab trade into the account (1841, 59–60):

> Adam was placed on mount ez-Zahun in Ceylon; there were leaves with which he covered his body, and as they were dry, the wind carried them off, and dispersed them throughout India. It is said that the frequency of perfumes in India arises from these leaves, but some have a different opinion: God knows best. They say, hence are, aloes wood, the clove, madder, musk and other perfumes particular to India. In this mountain sparkle diamonds and other precious stones.

Al-Masudi named his treatise *Meadows of Gold and Mines of Gems*, and this may seem exaggerated, but his descriptions related to facts on the ground. The Island indeed had extraordinary riches in gems, especially in the southwest. Castor wrote about this trade more than 400 years earlier (McCrindle 1897, 367–68).

> … this same Sielediba then, placed as one may say, in the centre of the Indies and possessing the hyacinth [amethyst or sapphire gemstone] receives imports from all the seats of commerce and in turn exports to them, and is thus itself a great seat of commerce.

These riches came from the alluvial plains of the rivers that drained from Samanakuta. As Al Tifaschi explained in the 13th century, "winds and rains carry the corundum stones, which are picked up as pebbles from this place, also as alluvial deposits" ([1253] 1998, 94).

These riches were related in the most direct way to the Peak. The Kalu Ganga (Black River), which rises south of Samanakuta and flows westwards to the coast, has huge beds of semi-precious stones near Ratnapura, a Sanscrit name meaning "city of gems". Ratnapura is also the head of one of the two trails to the top of the Peak. Al Tifaschi wrote that the Peak itself appeared to be studded with gems: "It is said that if the sun shines on this mountain, it sparkles as the sun rays reflect off the corundum stones; this is called the lightning of the Rahun" (1998, 94).

The appearance of the Peak on a Turkish map in the 11th century shows the wide currency of this version of Adam's fate, the wondrous surroundings of the earth where he was sent and the Peak's importance to the sea trade. The map was the work of a Turkish grammarian, Al-Kashghari, between 1072 and 1077 CE. Turkic tribes were essential to the silk trade between Persia and China which began in the 3rd century, so Al-Kashghari drew on experienced sources when he placed a small red scar at the southern extent of India and named it "Djebel Sarandib". Hermann (1935) notes that Kashghari did not know that the site of Adam's arrival on earth was on an island, but he did know the legend, for he inscribed next to it: "Here Adam came down, let peace be upon him!" (1935, 24). The Kashghari map is only known from a 12th century copy, but Kaplony (2008) was able to distinguish the mapmaker's original drafting and text from the work of the copyist.

From this time on, knowledge of the Peak and theories about its origin developed together with the mercantile economy, and this ensured an increasingly complex mix of indigenous and outside influences in its interpretation and use. Cosmas wrote in his *Topography* of an expatriate community of Christians among the traders, served by a "church of Persian Christians who have settled there, and a Presbyter who is appointed from Persia, and a Deacon and a complete ecclesiastical ritual" (McCrindle 1897, 363–64).

This community, or one like it, was probably the source of a Christian alternative to the account of the Buddha's footprint, a belief that Thomas the Apostle left it during his travels as a missionary. Thomas is only known to have traveled as far as Kerala in South India, and he was killed in Chennai (Hosten 1936). And 1000 years later, Fra Mauro, a Camaldulian monk who lived near the mercantile state of Venice, included the Peak on his Mappamundi (1457–1459 CE), the most detailed map of the world to that time. The special significance of the Island as a whole is clear on this map. The southern tip of India appears as two blunt peninsulas, but the Island is faithful to its actual shape, along with the position of the Peak – showing that it was much more familiar to navigators.

The footprint

Assessing the footprint is a different matter. It began as an idea, a resemblance in a naturally eroded basin in the rock on the summit, and as the Mahavamsa records, the Peak was already a holy place before the Buddhists recognized the footprint, the domain of the god Saman.

Belief does not want too much detail. While the earliest descriptions differed on the nature of the strange sight on the summit, they all agreed that it was the mark of a sacred being, and that was sufficient. Thus, physical descriptions are few, and they vary greatly.

Chao Ju-kua, known only from a single record as the Inspector of Foreign Trade in Fujian province in the mid-13th century, compiled information from the many foreign traders he encountered in a treatise he called *Chu-fan-chi* (Description of the Barbarous Peoples). In it, he offers the first detailed description of the Peak from the sea and from its summit (Hirth and Rockhill 1911, 73).

> There is (in this country of Si-lan) a mountain called Si-lun-tié, on the top of which there is a huge imprint of a man's foot, over seven feet long, and a like imprint is visible in the water (of the sea) within a distance of over 300 li [over 70 nautical miles] from the mountain. The forest trees on the mountain, little and big, all bend towards it (as if reverencing it).

Then, in 1344, Ibn Batuta, in his travels to China, stopped there, climbed to the summit and gave the first eyewitness account (Ross and Power 1929, 259).

> The blessed Foot sank into the rock far enough to leave its impression hollowed out. It is eleven spans long [about eight feet]. In ancient days the Chinese came here and cut out of the rock the mark of the great toe and the adjoining parts. They put this in a temple at Zaytun [modern Quanzhou, China], where it is visited by men from the farthest parts of the land. In the rock where the Foot is there are nine holes cut out, in which the infidel pilgrims place offerings of gold, precious stones, and jewels.

It was almost 500 years before visitors once again described the footprint. John Davy climbed the Peak (1818, 27) and saw, as Ibn Batuta did, that the footprint was now a sculptured object.

> … at present there are evident marks of design about it. I could observe on the surface traces of the labour of the workman: and the partitions which are between the toes, though resembling the native rock exactly in appearance, I found, on examination, to be a composition of lime and sand.

And in 1849, Charles Pridham describes the shrine in detail (1849, 615):

> This venerated memorial is five feet seven inches in length, two feet seven inches in breadth, and the slight similitude it bears to the shape of a foot, is produced by a margin of plaster coloured to imitate the rock: it is upon this moulding that the yellow metal case, which is profusely ornamented with gems of plain and coloured glass, is fitted before the usual time of the pilgrims' arrival. A temple built of wood surmounts the rock, and is kept in its position only by several strong iron chains fastened to the stone and also to the trees which grow on the steep sides of the cone. The roof is lined with coloured cloths, and its margin decked with flowers and streamers. This wooden temple, three feet high, dedicated to Saman; a

> pansala (priest's house), six feet square, built of mud; one large and one small bell (the former cracked), complete the catalogue of objects discoverable on the summit.

Even though Ibn Batuta accepted the footprint as Adam's and concluded that the Chinese had modified it, it seems clear from these and later descriptions that the natural impression was altered to fit the conventional style of Buddha's footprints.

Only in Ibn Batuta's passing mention of a toeless impression do we get a sense of what the original rock might have looked like, in its natural place on a small plateau of rock at the summit of this extraordinary Peak. In trying to imagine this experience, it might seem sufficient for confirmation and inspiration, especially among societies that worshipped the marks of supernatural beings in their lands. Applying the stereotype of the "other", the need for description and detail is the logic of strangers. However, this was clearly not the case here.

Instead of an ideological approach, consider the situation on the ground, the actual encounter of a pilgrim who had completed the difficult climb to the summit and walked up to the sacred rock. The act of cutting into the surface or augmenting the shape with cement to create a visible footprint ended the need for imagination. The print was now joined to the many others associated with the long and very complex tradition of the worship of padas (footprints) and padukas (sandals) of Rama and other sacred Hindu figures and of the Buddha (see, e.g., Kinnard 2000). And so, the sight of it directly reinforced the belief that brought the pilgrim on this long journey to the mountain.

One should not assume that the unmodified feature, even if accepted as an actual footprint, was sacrosanct. In Theravada Buddhism, natural features held to be actual prints of the Buddha as well as uddesika, objects or images that memorialized the Buddha, were objects of veneration, and so questions of authenticity are not usually relevant (Patterson 2012). However, since this particular footprint has such profound significance in Sri Lanka, it may be that an extraordinary circumstance provoked the modification. Looking back over the centuries, the work may have been part of the extensive efforts by Vijayabahu I in the 11th century to restore and reinforce Buddhism after the long occupation during the Chola Dynasty. Or it may have been earlier. The Mahavamsa chronicle rationalized and validated the Sinhalese migration and the ascent of Buddhism on the Island with a complex of spiritual, cultural and geographical logic, and it all began when Gautama Buddha placed his left footprint there.

With the deep relationship of Hinduism and Buddhism, and the fluid mix of diverse cultures in the region, the Peak was always there, with its dominant salience, to reinforce specific views of the world. Whenever the natural footprint was modified, it may have been, as Strong (2004, 95) suggests about footprint images in other places, "an imposition of Buddhist Order on a land".

Thus, more than 1500 years after the writing of the Mahavamsa, the Peak remains what the chronicle recorded and religious scholars pondered along the ancient sea trade route, the place where a supernatural being came down and left a footprint on its summit. What has changed is its original use as a navigation beacon, guiding traders across the open sea. By the 15th century, the rise of European mercantile nations ended the Arab domination of the trading routes, and with continuing advances in marine technology, reckoning by sea became less important, and so the Peak lost its status as a landmark and settled into its greatest role, as a sacred place.

This might seem to conclude a discussion of the significance of Samanakuta, but there is one odd thing about the whole of this history. Over all the centuries of wonder and worship that the Peak has inspired, there is a missing narrative – one that relates to what the Buddha found when he first traveled to the island: that it was already inhabited.

A forgotten story

There is always silence and space in narrative, when some words are not written, and when there are no words at all. At the beginning of the Mahavamsa in the first visit of the Buddha to the island to "win Lanka for the faith" (Geiger 1912, 3), his first task was to drive away the beings who lived there, the Yakkhas. The Buddha spreads his rug out with its edges on fire, pushing the Yakkhas away, and causes "the pleasant Giridipa [giri, meaning highlands] to come near to them, and when they had settled there, he made it return to its former place" (1912, 3–4). The chronicle continues on to relate the story of the founding of the Sinhalese people, with the arrival of Prince Vijaya and his followers to the Island, generally thought to be in the fifth century BCE.

Curiously, however, the chronicle does not abandon the Yakkhas to an unhappy fate. Instead, it appropriates them. When Prince Vijaya arrives, he quickly takes on a Yakkha woman as a bride and they have two children. Eventually, however, as the Sinhalese migrants settle, the Prince's advisers convince him to become king and take on a suitable queen. He woos the daughter of the Pandu king in Madhura, and at her arrival on the Island, his Yakkha wife is forced to go back to her people, is summarily killed and her two children flee "to the Sumanakuta", where, the chronicle concludes, they were the genesis of the "Pulinda" (1912, 67–68).

Outside the reference in the Mahavamsa, however, there are only literary references to indigenous inhabitants around the flanks of the Peak. In a Sinhala bird-messenger poem, the Parevisandēśaya (Message of the Dove), by the 16th century monastic poet Śrī Rāhula (Berkwitz 2017), John Bailey (1863) notes that the dove flies over the districts below Adam's Peak, and the poet writes that it will find there "'the daughters of the Veddahs' clothed in Riti bark, their hair adorned with peacock's plumes" (1863, 313).

Who were the Yakkhas, Nagas and Pulinda? It is curious that the Chinese Buddhist traveler, Fu-Hsien, who must have been enormously practical to travel so widely, could report matter-of-factly that before the coming of the Buddha, "The country originally had no human inhabitants, but was occupied only by spirits and nagas" (Legge 1886, 101). Nagas were long associated with water and serpents, and Coomaraswamy discusses Yakksas as tutelary deities (1928, 4). In general, these epithets were typically used to describe forms of otherness, whether strange creatures or foreigners. Pulinda was a common name for hunting and gathering tribes in southern India and Sri Lanka (Thapar 1971, 141). And the identity of the beings that Fu-Hsien described is actually clear in his own words, when he writes that these non-humans conducted silent trades with "merchants of various countries"(Legge 1886, 101).

> When the trafficking was taking place, the spirits did not show themselves. They simply set forth their precious commodities, with labels of the price attached to them; while the merchants made their purchases according to the price; and took the things away.

More than 1000 years later, the Englishman Robert Knox, who spent 19 years as a captive of King Rajasinghe II of the highland kingdom of Kandy, also describes this silent trade with the "wild men of the forest" (1681, 76–77). His account reveals the "Vaddahs" as a people in transition, with some groups still living by hunting and gathering and others at the margins of Sinhalese settlements practising horticulture. The conclusion is clear, as Raghavan writes: "Whatever may have been the forms or the characters of the Yakkahs and Nagas as we find them in the legends, we may accept them as representing the real, indigenous peoples of Lanka" (1964, 11).

So here in the founding legends of the Sinhalese people, the chroniclers found a way to resolve the awkward social reality of an incoming culture overwhelming an indigenous one. They first characterized the original inhabitants as non-humans, then joined into their indigenous origins through intermarriage and finally left them in the wilderness of the highlands, at the margins of society, and in the past.

Whether the early references describe the situation around the Peak is unknown, but the presence of the Vedda is implied in a much more fundamental way in the region. The old district and modern province of Sabaragamuwa, the area around the Peak, derives from *sabara*, a name commonly attached to tribal people (Thapar 1971, 18). And recent DNA analysis confirms the presence of a distinct indigenous population on the island before the Sinhalese migration (see Ranaweera et al. 2013).

With so few references, it is impossible to tell a story of these groups, to fill in the narrative, from histories that were obviously going to be more concerned with their own peoples than the ones they replaced. What is clear, however, is that for more than 2000 years the Vedda must have endured the rise and fall of dynasties and kingdoms, ethnicities and religions, accommodations and repression that spilled over from the mainland.

It is not simply the absence of records; with the great fluidity of cultures and beliefs on the Island over the millennia, no equitable history could possibly exist. So to help fill in the story of these ancient people, who inhabited the area surrounding the Peak, we need to look away from existing history and culture and consider how this whole environment, of sea, environment and human society, took shape.

An indigenous history

If there is to be a story of the Vedda that looks back into the past, it should resemble the Mahavamsa and have some sense of the origins of the world, also expressed in metaphor, and account for the shaping of the Island, the Peak, its human settlement and its position in a larger world.

About a hundred million years ago, the supercontinent Gondwana slowly cracked and began to open like a closed fist as the tectonic plates began to drift apart, and they now form a coastline running from eastern Africa to the Indonesian archipelago. The Deccan plate, comprising India and the island of Sri Lanka, was a discrete feature in the midst of Gondwana and it kept its central position, so that Sri Lanka became the most southerly landfall above open seas that run south to the Antarctic.

Within the island's shores, tectonic forces associated with the drift caused uplifts in the interior, forming a central massif of mountains and highlands, dominated by metaphoric rocks, surrounded by broad coastal plains. Updrafts in the high mountains lift water vapour into cooler skies where it falls back as rain, and with the added forces of monsoons, especially the northeasterly May to November winds that strike the massif on its southwestern side, downfalls are heavy and persistent. This drenching has weathered the mountains – in one place on the southwest edge of the massif, sharpening an isolated peak that rises steeply above its range – and it cut the mountain slopes into steep valleys, gathering the rivulets and streams into rivers that run through the lowlands to the sea.

As they run through the lowlands, the rivers drop these captured materials, forming deep beds of alluvium with scattered rock crystals of corundum and other hard minerals that had formed in voids in the mountain rocks during the immense heat and pressure of their formation.

At the beginning of human history, the Island was near the center of a great continental arch over the vast southern ocean, with littoral and interior environments offering wide variations in climate and biome, and along the ecotones, the potential for great diversity in resources. Even during glacial periods, the position of the Island near the equator ensured a favourable climate, especially in the southwestern forests, where the abundant moisture sustained a rich variety of plant and animal life through the last glacial maximum on the Indian subcontinent (21,000–19,000 BCE).

The first humans to live in these rainforests probably came from the west along the coastal corridor, which was likely a primary dispersal route out of Africa, during the Pleistocene (Roberts, Boivin & Petraglia 2015). Evidence shows that some were attracted to rock shelters in the southwest highlands surrounding the Peak. The succession of living floors in Batadomba-lena, Fa Hien-lena, and Balangoda Kuragala reveal occupation beginning about 30,000 BCE. From radiocarbon dates on charcoal (Kennedy & Deraniyagala 1989; Deraniyagala 2007), habitation intensified as the climate and biome moderated after 16,000 BCE. The sequence ends about 8000 years ago, as more recent activity in the caves and shelters has destroyed the shallow subsurface. The foraging life would have continued in much the same way in the early Holocene, however, as this area had an aseasonal climate and stable ecology following the glacial recession. For a considerable time, it was assumed that rainforests were barriers to early occupation (see Roberts and Petraglia 2015). Notwithstanding the occupation evidence that shows continuous occupation for millennia in the caves around the Peak, the analysis of tooth enamel from the three sites indicates a strong human reliance on tropical forest resources from 36,000 to 29,000 cal. BP through periods of evident environmental change at the Last Glacial Maximum and upon the arrival of agriculture (Roberts, Perera et al. 2015).

What this archaeological evidence suggests, therefore, is that by the 1st millennium BCE, when agrarian peoples began to migrate to the Island along the coastal plains and hills, hunting and gathering people had lived there for millennia.

This account merges with the historical reckoning of the Pali chronicles (Deraniyagala 2007, 9) and with historic eyewitness accounts. By the time of the Mahavamsa, horticulturalists from the mainland were well established in the dry coastal plains below the highlands. Cultural material of an iron-using society appears fully developed in stratified deposits at Anuradhapura in north-central Sri Lanka, ca. 900 BCE, and writing appears in subsequent strata in an Indo-Arayan Prakrit dated to ca. 500 BCE (Deraniyagala 2007, 9). And when these incomers gained the technology to construct the reservoirs and storage tanks they were able to use more intensive swidden and rice cultivation technologies (Morrison In Press, 238).

The foragers would have continued to live in their traditional ways in the forests, but the forest–plain ecotone is a porous boundary, as it affords much more diversity in resources than either environment on its own – and forest and coastal groups would have exploited both. Unfortunately, the favourable ecological conditions were countered by a severe imbalance of essential resources when it came to trade and exchange. The highlands and coastal forests, especially in the southwest, were rich sources of spices, aromatics, exotic woods and gemstones, and as Cosmas observed (McCrindle 1897), they were already well exploited at the time of the chronicles. The forest peoples were at least initially involved, as the nagas in Fu-Hsien's 5th century description traded in "precious commodities"(Legge 1886, 101).

Eventually, the slow creep of horticultural settlement in the dry zones of the north became a broad advance into all but the remotest uplands. As the settlers moved southwards they cleared forests, using *chena* (swidden, or slash-and-burn cultivation) in the lowlands of the wet zone (Broun 1900, 355), and reached the Kandian kingdom in the central highlands in the 14th century.

When traditional trade was replaced by colonial exploitation, the old-growth forests near the Peak were largely stripped in the search for the ebony and other exotic woods that thrived in the wet zone. By the 17th century, the Island was fully involved in international trade and European colonists had begun to appropriate territory on the coasts. In Kandy, Knox (1681) witnessed what he described as "tame" Sinhala-speaking Vaddahs living on the margins of Sinhalese society and "wilder" groups living in the forests. Some groups simply *chose* to resist this change in adaptation, and so remained in the highlands, away from agricultural settlements, where they could pursue their traditional ways. These foragers continued to live there well into the 19th century; Bailey observed they were still "at home" in their hunting grounds in eastern coastal regions (1863, 285).

The problem with most sources relates to the prevailing attitude that hunting and gathering societies were the survivors of declining and primitive lifeways, rather than groups with social integrity able to adapt to the situations at hand. This was evident in the 17th century, from Knox's descriptions, which show a society that adapted to introduced technology without surrendering the old (1681, 64).

> They are so curious of their Arrows that no Smith can please them; The King once to gratifie them for a great Present they brought him, gave all of them of his best made Arrow-blades: which nevertheless would not please their humour. For they went all of them to a Rock by a River and ground them into another form.

More than 200 years later, Seligman and Seligman (1911, 422), in their detailed study of the Vedda, concluded that "Perhaps the few surviving 'wild' Veddas have altered less socially than the people of the [so-called Dravidian] Indian jungle groups" – a sign of cultural integrity rather than cultural decline, even as the population declined.

On the flanks of the Peak, it was a different story. The exploitation of the southwest highland forests turned into devastation after the late 1860s, when the first tea was planted. By 1900, Sabaragamuwa Province had only isolated blocks of forest, protected by their inaccessibility, and the loss of these old-growth forests caused severe erosion and flooding in the lowlands (Broun 1900). With the destruction of the forests, traditional foragers had to either move northwards into the interior, or adapt to the dominant horticulture society – if that was possible. Percival (1803) noted that groups in the southwest were more assimilated than groups north of Kandy, suggesting that many chose to stay and adapt.

There are no further references to indigenous groups near the Peak. The Chronicles move on to the history of the Sinhalese, the Peak becomes the great mountain with Buddha's footprint and the foragers who once lived on its flanks are eventually overtaken by noteworthy events and forgotten. Yet, Bailey (1863) refers several times to Vedda spiritual beliefs and activities associated with rock outcrops and hills, which suggests that they would have held the Peak in highest regard. It is therefore probable that before the arrival of the Sinhalese migrants, they lived in the Peak's surroundings and also looked to its summit with wonder and spiritual regard. Perhaps modern descendants of the Vedda should have some better part of the narratives of the Peak than they do now.

The deeper surroundings

If we add the germ of a story of the Vedda to the narrative, it enlarges the already powerful ecumenical presence at the Peak, and we might be satisfied about such a history – but there is still one question to be answered. What is it about this mountain that began to attract people even before the knowledge of the footprint, and is so compelling that it even overcomes profound religious differences?

The question is difficult because what we see rising about the highlands is the Peak. This pinnacle is a governing representation, a cumulation of all the narratives that make up its history, which provides the descriptions for interpretations such as this. Indeed, as humans exist through description, the Peak we know is the cumulation of metaphorical resources from the speech, writings and actions of inhabitants, pilgrims, explorers, traders, colonists, tourists and analysts over millennia. And, like all great narratives, it absorbs the nature around it as if the whole world is part of its human preoccupations. At the onset of the monsoon in January, for example, the Peak takes on a special name, *Samanalakanda*. Great clouds of butterflies begin to migrate towards the highlands and the Peak, and people believe that these great swarms are pilgrims on their way to the worship at Sri Pada (Woodhouse 1950).

The problem is that if we only think about the meaning of the Peak and its footprint, we miss another whole world – that of the mountain, along with the direct, sensory experience of it. As this world exists within the forces of nature, it has its own imaginative history, focusing on the mountain in a live environment, and on the physical affordances that underlie its extraordinary social effects.

This history, if written, would begin with the accidental conjunction of material and social affordances related to the Island's place in the world. In the lands that would become Sri Lanka, the unique confluence of minerals and tectonic forces produced rocks studded with crystals; the drifting of the tectonic plates gave it a central position along the arch above the vast southern ocean; and its proximity to the equator and a mountainous interior resulted in the growth of very warm, aseasonal rainforests in the southwest, above the broad coastline and the sea. After millions of years of weathering and natural life in this Island, flows of people, technologies and materials eventually extended into these highland rainforests. As an oddly shaped, conical form, isolated from other peaks, it commanded attention and so guided coastal migrants through the deep forests to its flanks, where some began to settle and, eventually, it attracted ships at sea. It was not only the visual impact of this pinnacle that gave it such extraordinary power. The rainforests and sedimentary plains below the mountain are unusually dense in sensory resources – the gemstones, aromatic plants and exotic woods – and these would eventually dominate the Island's economic and cultural life. All this accounts, therefore, for the impact of the Peak as the grounds of a deeply shared environment that crosses time and society.

The confluence of physical and social flows at the Peak show that its drawing power is more fundamental than it is possible to articulate. If the task is to understand its great presence, it is clearly necessary to face this much sparer narrative and take into account the physical world at work there. Consider the direct experience of the pilgrims – individuals, families, young and old – who choose to climb the difficult pilgrimage trails. Surely there is more to their motivation than following the prescriptions of belief. When the pilgrims walk where the Buddha once walked, pausing to rest at the sacred place where it is said he paused to sew a tear in his garment, or to have a drink of spring water, what may appear to be communal action is a confluence of urges, desires and capacities in each individual, along with a striving that comes in part from sharing the task with others. Such experience is not drawn down from above. It resides in the experience of *being there*, of seeing and climbing more than 5000 stone steps upwards, towards something in all ways outside ordinary life.

The lure of sheer experience is evident in the desire of many pilgrims to climb at night, despite the dangers, because they hope to witness a stunning effect that sometimes occurs at sunrise. A few minutes before, if there is low cloud on the horizon, a special *ira sevaya* may take place (Abhayasingha 2012). At first, the sun seems to flare above the horizon, but this single arc suddenly divides into several pieces or moves from one place to another, until all the light gathers into a single, blazing ball. Abhayasingha interprets the effect as the total internal reflection and diffraction of light by the lower atmosphere at the distant horizon, while pilgrims and Buddhist monks see the sun worshipping the Buddha's footprint three times before it rises.

A second strange event may occur within minutes of the sun's rise. If the sky is clear and mist still covers the hills and plains below, the Peak may cast a remarkable shadow across the highlands. In Skeen's (1870, 222) description,

> … as the sunlight became stronger, the outline and body of the gigantic pyramid shaped umbra grew sharper, darker, and more distinct; and as the sun rose higher in the heavens, the titanic shadow seemed actually to rise in the atmosphere; to tilt up and gradually fall back upon the

> mountain, shrinking and dwarfing in dimensions as it drew closer and yet closer to its mighty parent, until, absorbed in the forests with which the mountain is clad, it was wholly lost to view. So singular a sight, — one so strangely magnificent, and even awe-inspiring, can be seen nowhere else in the Island, perhaps nowhere else in the world.

The joining of these natural and cultural forces, at what seems there to be the highest point in the world, radiates a power directly to the senses that exists only until these moments pass – perhaps by the ring of a bell, or someone's exclamation – and then awareness and social selves return, the experience becomes a reflection and it appears as if the gods had ordered such things.

The senses seem lost in the dense ideological surroundings of the summit, yet, if we treat the mountain as the host of its world, a place that people may attend without the need for instruction, it may be possible to open up analysis to what does not fit into the narrative picture – pilgrims who fail to make it to the top, the people in all the small settlements below, the indigenous Vedda, who are not included at all, and the extraordinary natural forces involved on the slopes and summit.

In a practical sense, it is difficult to see Samanakuta in this way, outside its narratives. More than 2000 years of belief and tradition have overwhelmed memories of its ancient past, records are sparse, the indigenous inhabitants are gone and the possibilities of further exploration seem limited. In the Black Hills of South Dakota, however, there is another great pinnacle that has attracted people from many cultures over the millennia to its flanks in remarkably similar ways for at least 10,000 years, and it too has inspired a host of origin stories and spiritual traditions. Like Samanakuta, the power in this formation is a confluence of the way it strikes the eye and the imagination, its consistent and stable nature and its extraordinary features. This rock has no meaning in itself, but neither is it simply a placeholder for stories that illuminate beliefs and spiritual life, or scholarly study. To gain some insight into the dynamic world that monuments afford, we will return there and consider the evidence of how some of the indigenous groups of the region viewed, pondered and used it.

Tso-Aa, He Hota Paha, or Devils Tower

The sight of the Tower

If we look from the perspective of this Tower, which stands at the edge of a sea of grass, we will begin this narrative by looking at its geology and geomorphology – the natural history of its affordances. This provides the necessary grounding for the discussion to come, about the field of possibilities in its surroundings that account for its enduring significance.

The tectonic forces that forced up the Rocky Mountain chain in western North America between 80 and 40 million years ago also produced a small, high dome on its eastern frontier. The new mountains drew rains out of the weather fronts moving from the west and the resulting erosion carried the uplifted sediments eastwards until they eventually pushed the water back, leaving the dome as an island in a vast sedimentary plain. Its erosion produced the unusual terrain of exposed core rocks and imbricated ridges that we know now as the Black Hills.

To understand the settlement history of any region, we have to look at its water resources and their flow dynamics. In the Black Hills ecosystem, the pattern of rivers and streams is the result of much more recent geomorphological changes. The Rocky Mountain's front range streams once drained northeast towards the Arctic Ocean, but during the Late Pleistocene glaciation, expanding ice sheets in the Canadian Shield forced drainages eastwards and then abruptly south along the western side of a lobe that covered the east half of present-day central South Dakota. This postglacial Missouri River gathered all the east-flowing drainages in its basin, eventually draining south into the Gulf of Mexico.

As the glaciers retreated in the Late Pleistocene, about 140,000 years ago, grasslands replaced the cool parklands in river basins and uplands, forests and meadows covered adjoining mountains and animal life flourished; and streams flowed from the front range of Rocky Mountains, through the High Plains and around the Black Hills and then down through easterly grasslands to the distant Missouri.

The easily accessible Black Hills ecotone attracted foragers who moved northwards at the end of the Pleistocene. Small occupation and food processing sites are scattered through the Hills, most often at streams and springs and in the surrounding plains. A hunter among one of these groups used the projectile point that fragmented in the gulch on Mt. Rushmore. The few cultural sites dating to the Late Pleistocene and early Holocene lends to the whole the feeling of a frontier, but this may be related to the limited evidence in the archaeological record. Cultural sites with long-term occupations, such as Hell Gap, about 85 miles southwest of the southern edge of the Black Hills, show that the foragers who reoccupied lands around the Black Hills uplift after the glacial recession were skilled toolmakers and probably flourished in the region. And while no rock art of this time has yet been dated, the Powars II ochre mine at Sunrise, Wyoming, about 6 miles from Hell Gap, was active 13,000 years ago, indicating a demand for the pigment (see Tankersley et al. 1995). The small foraging groups that followed a seasonal round among the Black Hills and surrounding grasslands probably had a cultural life as full and complex as the late Pleistocene and early Holocene peoples we know best from caves in other parts of the world.

The big picture of millions of years of geology seems to lead to a big picture of human history, but this is partly an illusion. The geological history was great irruptions over vast stretches of the earth, continents building up and eroding, and the life that followed, while the marks of humanity upon it are tiny. Most often, this human past shows in the time and space of a brief occupation: the lens of ash, charcoal and cultural debris of individual hearths, surrounded by scatters of debitage, tools or fragments of animal bone, or images cut or painted in an hour or two on a rock face. These finds simply trace ordinary life, preserved in its own grounds.

With two visions, different in scales of time and effect, and with so little evidence of occupation to go on, the only way to reconcile them is in an overview. In the usual routine, collections of site data become generalized cultural and social traditions, and these constructs, remote from individual sites and situations of production, stand in for the individual cases. Thus, we identified the ancient hunter in the gulch because the projectile point fragment resembled types found at sites about 9000 or 10,000 years ago, and thus indicating membership in one of the widely distributed social groups active at the time.

When the norm is creating an identity for individuals and social groups through their sites, it may seem pointless to go back to questions concerning the actions of individual people on actual ground, foraging, hunting, living in their camps and moving through their familiar surroundings. A projectile point fragment in the gulch is not much to build a society on. Yet, each individual site is important because the artifacts take shape at this human scale, the discrete interaction between living individuals and live environments – in the case of our fragment, on a ledge above a stream that still runs. As streams and gulches persist, so do the people who follow them and, while they all think differently, they stand on common ground – the ground, of course, on which the archaeologist also walks. The tacit knowledge of a territorial bond is what motivates archaeologists to survey areas by the possibilities of the physical terrain and resources, as most sites, it turns out, seem just as appropriate for occupation today.

For this reason, in trying to understand the presence of a great landmark, a pinnacle at the edge of the Black Hills, we have to consider the encounter in the course of daily life that influenced all the people who came by, walking similar walks.

Imagine the circumstances in a trek to the Black Hills on foot from some distant point in the plains – the experience of most people before the introduction of the horse to this country in the mid-18th century. Rolling grasslands flow to the horizon like the sea, broken by scatters of buttes and hills, and the forks and meanders of river valleys seem endless. Yet, there is no need to invoke cognitive maps or landscapes to imagine how people found their way. Two streams rise east of the Rocky Mountains, flow up to the Hills and divert around them, one on the north, the Belle Fourche, and the other on the south, the Cheyenne. They eventually converge in the Great Plains as the Cheyenne, and drain into the Missouri more than 300 miles to the east.

Following either of these streams on foot is not straightforward. They always meander, and their valleys may be steep and choked with trees and thickets, so travelers will try to skirt them in places, cutting across hills and ridges, or crossing in shallow water. Out of the river valleys, navigation in the grassland sea is by buttes and hills, while at night, the stars define the general path. Eventually, whatever the direction of the journey, the dark forests in the Hills appear on the horizon.

Through such a journey downstream, we come to the place of our interest. With the Black Hills in sight, the Belle Fourche flows northeast through an arid plain, scattered with cactus and yucca, and then winds through the Bearlodge Mountains, an outlier of the Black Hills on the north. Near the western edge of these mountains, three oddly rounded hills rise out of the plain, and soon after, when it is possible to look further downstream, an exceedingly strange object comes into view. It is an enormous stone monolith with straight, ribbed sides and a blunt top. It rises more than 1200 feet from a huge mound of sediment and shattered rock, ringed by sandstone cliffs, at the edge of the river.

Captain William F. Raynolds, a cartographic engineer, was the first to record his impression, writing on July 18, 1859, that he spied the rock "rising like an enormous tower" far in the distance. His military exploring expedition was following the Cheyenne River west of the Hills through Lakota territory to the Yellowstone (Raynolds 1868, 32). His topographer, J.T. Hutton, and Zephyr Rencontre, a French fur trader and Sioux interpreter, reached it two days later. Raynolds seemed oddly dismissive: "it is, as I had supposed, an isolated rock upon the bank of the river, striking only from the fact that it rises in a valley" (1868, 32). For him it was a point on a map, but had no strategic value at a time when the United States was exploring and mapping its recently acquired territory. How different it was for the indigenous hunters and gatherers who had walked along these same routes over thousands of years.

Almost 30 years later, in 1897, an elderly Kiowa man, I-see-o (Plenty Campfire Places), who had served as a scout for the U.S. Army, shared his knowledge of the Tower with his commandant, Major General Hugh L. Scott at Ft. Sill, Oklahoma (Scott 1920, 163). He never learned to speak fluent English, so he conversed in a way that both he and Scott were fluent, in Plains Sign Language (e.g., Davis 2013). From this exchange, Scott wrote his account of I-see-o's story.

> Before the Kiowa came south they were camped on a stream in the far North where there were a great many bears—many of them. One day seven little girls were playing at a distance from the village and were chased by some bears. The girls ran toward the village and the bears were just about to catch them, when they jumped up on a low rock—about 3 ft. high—one of them prayed to the rock, "Rock, take pity on us—Rock save us"—and it heard them—it began to elongate itself upward, pushing the children higher and higher. When the bears jumped at them they scratched the rock, broke their claws and fell down on the ground—the rock rose higher and higher, the bears still jumping at them, until the children were pushed up into the sky, where they now are, seven little stars in a group (the Pleiades). In winter when they are just overhead it is the middle of the night. The marks of the bears' claws are there yet—just

> like the side of the "Medicine Bluff" (Townsite, Ok, near Fort Sill, a massive rhyolite cliff, riven with vertical seams). No Kiowa alive now has ever seen that rock, but the old men have told about it—it is very far in the North where the Kiowa used to live—it is a single rock with scratched sides (rising up straight to a great height). There is no other like it in the whole country—there are no trees on it only grass on top. The Kiowa call this rock "Tso-aa" a "Tree Rock", possibly because it grew up tall like a tree. When they came to look they found bear claws all around the base turned to stone.

This detailed account combines an origin story, physical description and history, and it is remarkable because, as I-see-o relates, "no Kiowa alive has seen that rock".

This was not always the case. The Kiowa once ranged so widely that their language relates to the Pueblo of the Southwest, and more recent oral tradition and historic records trace a migration southeast from the Cascades in western Montana and British Columbia southeast into the High Plains (Meadows 2016) and, at its western extents, the Black Hills. By about the middle of the 17th century, they had obtained horses, and this radically increased their mobility, as well as their ability to hunt bison across the plains (Gunnerson and Gunnerson 1971).

The Kiowa were not alone. Arapaho, Cheyenne, Crow, Lakota and Eastern Shoshone all adapted to this new economy and included the Black Hills in their homelands and hunting grounds. And this sudden facility to travel caused much cultural change. For example, Shimkin observed that when Eastern Shoshone bands moved from the Great Basin to the Wind River mountains at the edge of the High Plains to take up bison hunting, it "exposed them to many streams of influence, contributed to a great fluidity of culture and choice of many cultural alternatives" (1942, 461).

There was continuous manoeuvring and realignments of tribal societies in the Plains as the 18th century progressed, with considerable conflict. This was largely an effect of what was happening along the east coast. The French and British, competing over control of the fur trade, facilitated the expansion of the Iroquois and the Ojibwa westwards, and the Sioux tribes, now mounted, began to move from south of the Great Lakes into the Plains. Under pressure from the Sioux, the Kiowa began to move from the Black Hills area southwards and they eventually settled in the southwestern plains of Oklahoma. And by the time that I-see-o told his story, it was more than 100 years since his people had left their northern homeland.

The Tower had this effect on all who saw it. It figures in the oral traditions and histories of virtually all the Tribes who encountered it, despite differences in language and society.

This is consistent with Sundstrom's observation that when a new group settled in the region, they tended to adopt existing traditions, despite cultural differences, and modified old ones "to the new physical and conceptual landscapes in which groups found themselves" (1996, 188).

They could also lose this tradition when they left. Consider one Southern Arapaho story that follows the Kiowa version closely. This Algonkian-speaking tribe once lived by the Great Lakes, moved west in the 18th century to hunt bison and eventually settled in the southern plains of present-day Kansas and Colorado. They were involved in the competition for lands that pushed the Kiowa south, but by 1840 they had allied with them. In this account, the giant bear chases seven children, but here there is no tree rock; the heroic actions of one girl sends them all into the sky as the star cluster, and in the end, the bear, thwarted by the children, scratches itself (Dorsey & Kroeber 1903, 152–53).

The difference between the Kiowa and Arapaho treatments may relate to the territorial bond each had in the Black Hills. The Tower was at the very edge of their traditional homeland, and so a story about the Tower may have moved and changed more easily. In contrast, the Kiowa left a life in the Black Hills, with camps, hunting and gathering grounds, burial sites and sacred places, and the Tower was an anchor for these memories.

This turn towards the ground and its topographical features, perceived and used in common, lends to the Tower a sense that it organizes the flows of people around it, rather than it being in the service of the people who find it – adaptation, not appropriation. Indeed, for all inhabitants from the deepest past to the present, and for all the narratives upon narratives that exist, there is only one Tower. In its river trench, on the northwest margin of the Black Hills, the Tower is a world on its own terms. Long before the first humans arrived, vultures and other raptors soared in the thermals around its summit, as they still do.

If we can imagine life from the position of the Tower, its obviously intrusive nature and peculiar shape give it an alien presence in otherwise familiar surroundings, and this effect affords practical behaviour and, on the evidence of oral traditions, beliefs and historic developments, provokes interpretation. This is evident in recorded speech about the Tower, often generalized as oral tradition related to what these stories might signify. If we treat the received narratives naively, at face value, in terms of the individuals who related these stories, we may be able to derive aspects of their experience of being there.

The shifting shape

There is only one Tower, but its shape depends on the direction from which travelers come. This is preserved in some of its names. A traveler following the river downstream approaches the formation from the west and southwest and from this direction (Figure 3.3), it looks like a huge tree that has snapped cleanly at the top.

This is Tso-aa, the tree rock of I-see-o's Kiowa story. Near the rock, the deeply furrowed trunk spreads out at the base as if it has a huge mass of roots. While I-see-o explains that the furrows are the claw-marks of the bear, there is in the name a deeper resonance that may relate to the tribe's origins. The Kiowa believe that they emerged from a hollow cottonwood log, somewhere in their northern lands (Mooney 1898, 152; Meadows 2016) and,

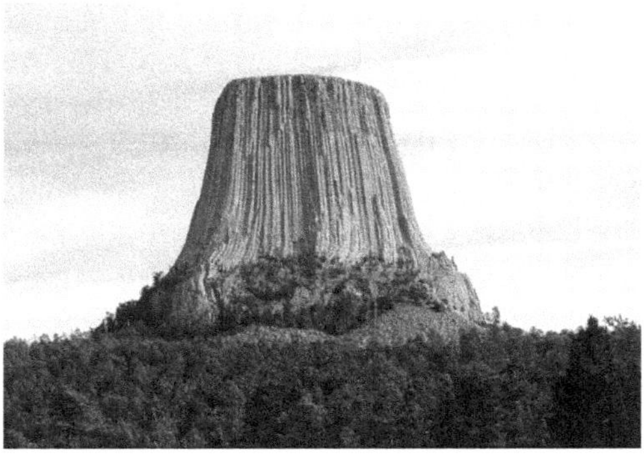

Figure 3.3 Devils Tower

indeed, the deep furrows closely resemble the bark of an old cottonwood tree – a familiar sight along the Belle Fourche River. This suggests a more complex background to the Kiowa story, which converges because Kiowa people were familiar with the Tower from their camps and in their travels, rather than recounted mainly in stories.

Another western tribe uses a name drawn directly from experience. In 1887–88 a small group of Shoshone traveled eastwards from their home in the Wind River mountains past the Black Hills and the Standing Rock Sioux Reservation in present-day North Dakota (Shimkin 1947b). One of the travelers, Pivo Brown, recalled this journey to Shimkin. His account only mentions the Tower in passing, and in translation; he uses the common English name: "We went all day, past Devil's Tower" (1947b, 30). On Shimkin's map of Shoshone homelands, however, Pivo Brown's route is marked with the Shoshone name, which translates as gourd-its-butte (Shimkin 1947a, 250). By this description, the ribbed sides are not the grooves of a tree or the gouging of bear's claws, but the stem of a gourd, a desert melon (e.g., *Cucurbita foetidissima*, the Buffalo Gourd). Once all the Shoshone lived in the semi-arid Great Basin, before the introduction of the horse enabled their western relatives to take up buffalo hunting in the plains west of the Big Horn Mountains and the Wind River range. The Shoshone are very protective of their history and religious traditions, so it is not known to what this refers, but it does suggest an historical reference to past life in the semi-deserts of the west.

For someone following the river upstream, approaching from the northeast, the straight-sided formation gives way to one that is distinctly concave on its western side, an effect accentuated by the sweeping arc of the furrows. This approach was common for Lakota people, who ranged westward through the northwestern grasslands from the Missouri River and hold traditional lands (and legal claims) in the Black Hills today.

In a recent placename map of Lakota traditional territory in western South Dakota (Lower Brule Sioux Tribe 2017), Lakota elders of the Kul Wicasa Oyate give its name as *He Hota Paha* (Gray Horn Butte). As a buffalo (*Bison*

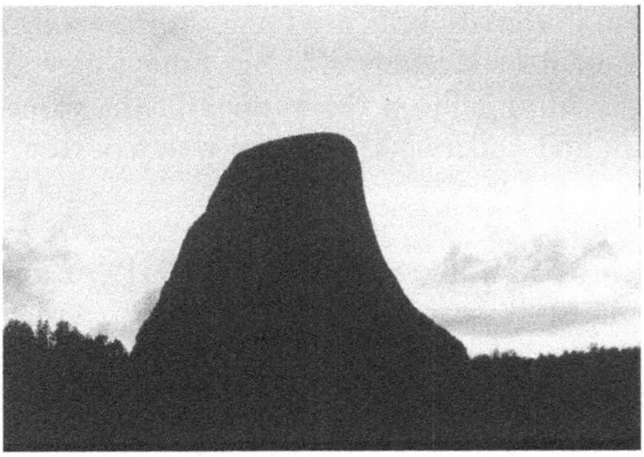

Figure 3.4 Devils Tower, View from the Northeast

bison) ages, its horns, once jet black with finely shaped points, begin to grey as the base grows upwards, wrinkled and scaled, and thickens (Hornaday 1899, 33–34), and they may end up blunt, battered and broken. And when the animal dies, the horn covering eventually falls away, leaving the hardened core exposed, flat and riven like the tower, where it might be discovered, sun-bleached on the prairie.

In contrast to unique names that describe the entire formation, which preserve in them the direction from which they were viewed, the most common names of the Tower account only for the furrows, and they are all variations on the incident in the Kiowa story, in which a giant bear claws at the sides of a moving rock in pursuit of children sheltering on its top. Another Lakota name for the tower, Mato Tipila (Bear Lodge), reflects this story.

As the Southern Arapaho example above shows, the bond between story and place may be fragile and, indeed, such plots and story elements appear elsewhere. Rocks or trees that suddenly grow, sometimes taking people into the sky, feature in myths across North America (e.g., Waterman 1914); and the bear has a worldwide mythological status, befitting the complex functional, psychological and social complexities of its relationship with humans, as Hallowell (1926) shows in his study of bear ceremonialism.

If we look at the bear story at the Tower as a mythological theme, however, it loses its grounding and, therefore, anything that might contribute to understanding the experience of the people who lived there and told the story. The analysis needs to make sense in terms of life in the Black Hills and the Bearlodge Mountains, and, most particularly, in relation to this strange rock in the valley of the Belle Fourche River.

As the stories indicate, many bears once lived there. Its habitat of mixed, open forests is suitable for grizzly bears (*Ursus arctos horribilis*) and black bears (*Ursus americanus*). Indeed, bears foraged well into the Great Plains. On February 16, 1806, Meriwether Lewis wrote that

> The brown white or grizly bear are found in the rocky mountains in the timbered parts of it or Westerly side but rarely; they are more common below the rocky Mountain on the borders of the plains where there are copses of brush and underwood near the watercourses.
>
> (Thwaites 1905, 78)

The grizzly was extirpated in the region in the late 19th century (Mattson and Merrill 2002) and the black bear had largely disappeared well before the end of the 20th century. At the time elders shared these stories with their families, however, bears were certainly a potential problem. Bears are curious and opportunistic scavengers who may prowl around camps. They seem lumbering and clumsy – but when they are foraging, black bears move very quietly for their size and, if there are no dogs about, they may only betray their presence at night by the snap of a twig. And both species may be dangerous if provoked – especially grizzlies, which will defend their young aggressively; and both grizzlies and black bears can charge at speeds of more than 30 miles per hour.

There is more to this relationship, however, than a stereotypical fear of wild animals. Distinct from all other larger carnivores, bears have plantigrade (flat) feet like humans, and can walk for short distances on their hind legs, and even sit upright, with their legs crossed (Hallowell 1926). And some societies see human features in the skinned bodies of bears (1926, 149) – even our own, as we noted above about the similarity of skeletal bear paws and human hands (Sims 2007).

The great bear's actions in the Kiowa story also came from familiarity. Males often rub, bite or claw at the side of a tree as a form of territorial marking and will use the same trees repeatedly – sometimes shredding the bark. Thus, it makes practical sense that a giant bear might claw at a tree rock as it chases a group of children.

Finally, the association of the origin stories and lived experience is also evident in the climax, where the children become stars in the night sky – in I-see-o's account, the Star Girls (Meadows 2008, 290). To contemplate this joining of earth and sky at the Tower, it is important to recognize our biased conceptions of the sky and outer space, which we owe to more than 2000 years of astronomy. For the cultural groups who can only know the skies through what they see with the naked eye, there is no way they could know that points of light in the sky are planets and stars turning in a dark void or how far away these lights might be. They might as well be beings moving about a distant, unreachable land, or light showing through a roof or dome. Where all studies of the stars and planets agree, however, is in their visibility and the predictability of their motion, which helps track days and seasons and guide journeys. It is therefore reasonable to see the night sky in any culture as a terrain as interesting as the lands within the horizons of the day. In the Black Hills area, Goodman (1992) has recorded a significant body of contemporary star knowledge among the Lakota, mainly through interviews with elders.

While the patterns of constellations are relative to the societies that define them, there is considerable convergence. Many cultures recognize that one star remains in its place in the northern sky (Polaris), and the patterns we know as the Big Dipper (Great Bear, or Plough) and Orion are found elsewhere. The Pleiades, however, stands out as a unique case. It is a star cluster, rather than an asterism or constellation, and because it is so compact, it stands out as a single feature. In an unpolluted sky, the seven brightest stars are readily visible to the naked eye and, with concentration, it is possible to view several more. Winnecke (1878) reports that in the 16th century before the telescope, the astronomer Michael Moestlin made an accurate drawing of 11 of its stars. This accounts for its wide distribution in place and time. Its pattern of seven stars appears on the Nebra Disk (see Pasztor 2015), dating to 1600 BCE, discovered in Germany, and it appears widely in indigenous art.

If the visibility of the Pleiades was a significant factor in its use, it does not follow that the original speakers of I-see-o's Kiowa story simply picked this star cluster from the sky to fit their narrative. For someone living by the Tower in one of the surrounding meadows, other possibilities present themselves. In the course of a day, the rock seems to change shape, as light and shadow move across its ribbed surface, and as night falls it is the sky that moves. Even during a moonless night, the dark mass of the Tower looms up into the light of the Milky Way.

In mid-November, the Pleiades are visible all night in the Great Plains, and someone who watched the stars from a position west of the Tower would see the Pleiades rise above the summit in the early evening, while east of the Tower they could watch as it set due west between the middle of the night and dawn. The Tower could thus act as a nocturnal gnomon, providing both spiritual and practical knowledge related to the seasons.

What is most important about the details in the Kiowa story and in all the other accounts discussed here is that they relate to experiences at the Tower, rather than to ideological dynamics. After I-see-o finishes telling Scott about the rock's origins, he briefly mentions what Kiowa people saw there to affirm the story – and this has one remarkable detail that shows clearly just how specific the Kiowa knowledge was. He concludes with this observation: "When they came to look they found bear claws all around the base turned to stone" (Scott 1920, 163). Only people who searched carefully around the edge of the Tower mound would have come upon small fossil beds eroding out of a small hill near the base of the formation (Molyneaux et al. 2000) and seen unusual stones that resembled broken claws or teeth. These are the sheaths, or guards, of a species of Belemnite, an extinct cephalopod of the Jurassic period related to squids (see Tweet 2019). So important are these stones to the Kiowa, that elders still travel to the Tower hoping to see them (see Meadows 2008, 82–83).

Taken as a whole, the narratives about the Tower may be consistent with oral traditions and folklore in themes and subjects widely distributed in the region, but some are also recollections of direct experience, which brings the stories back down from the skies and into the camps of people who lived there.

A place to live

The Tower stood above the surrounding sediments for millions of years before humans began to forage in the region following the last glacial recession. By this time, the prevailing winds and the constant cycle of freezing and thawing had shaped it into an irregular oval and battered the west face flat, resulting in the huge field of shattered columns and rubble that fans out from the flanks. On the sheltered east and southeast sides, there is much less talus. Here, relatively flat areas of open ground, part of a narrow collar of ancient sediments, have resisted erosion. Moving away from the Tower, the outer edge of the mound shoulders is a ring of sandstone cliffs and, below this, the ground slopes steeply to the edge of the river valley and the Belle Fourche River, which runs up to the formation on the southwest, meanders around the Tower hill and then continues northeast into the Bearlodge Mountains.

At the tower base, the impacts of wind and water have created a series of environmental niches, with much potential for occupation and other practical uses. The blocky talus on the west side looks like the debris from an avalanche; it is almost 150 feet deep in places and many blocks are huge and angular but, close to the Tower, small pockets of exposed soil provide good shelter. On the north and east sides, the sheltered open ground has well-developed soils that support trees, shrubs and grasses. This would have been an ideal place to camp. Where the talus forms a sheath around the lower elevations of the Tower, regolith and soil have also built up in niches and benches, and scattered Ponderosa pines and other vegetation cling there. On the blunted summit, grass, sagebrush and other plants grow, and snakes and chipmunks shelter in the fractures.

The archaeological evidence for occupation comes mainly from a reconnaissance survey of the two square miles of land around the Tower, conducted by the National Park Service, in what is now Devils Tower National Monument (Molyneaux et al. 2000; Molyneaux 2006). Since the monument was established in 1906, there had been sporadic surveys, but the dominant impression was that there was little to find there. After repeated finds of lithic material on the slopes, however, the NPS initiated a comprehensive survey to determine the nature, extent and distribution of cultural resources surrounding the Tower, for the purposes of protection and preservation.

The results showed that before the coming of Europeans, the Tower area was a focus of intense activity, typical of ordinary hunting and gathering life. The distribution of artifacts and materials suggests two types of occupation: one typical of any river valley in the region; and the other, focused on the affordances and resources at the Tower.

The river valley provided all the food, shelter and other resources necessary for seasonal occupation, and this probably accounts for the numerous small scatters of artifacts along the edge of the valley. Given how close these sites are to the Tower, it is hard to imagine that these travelers or local groups did not climb up to the Tower – but they did not need the resources of the Tower to live there.

This timeless pattern is not the same, however, for the sites that cluster around the high meadows and open talus on the summit of the Tower mound, which is at least two-thirds of a mile from the river and approximately 400 feet higher. These occupations took direct advantage of the Tower's unique affordances. The most obvious function relates to the spirituality implicit in the origin stories about the Tower. Many informants, Sioux, Arapaho, Cheyenne and Crow elders that Stone (n.d.) interviewed in the 1930s, said that they had camped there for vision quests and ceremonies.

With its great height above the river valley, it is inevitable that the Tower would also be a local landmark. What makes it remarkable, however, is the alien quality of its ribbed sides, which are so distinctive that travelers could carry its description far and wide, or travel from some distant place to the Black Hills and find it easily. The extent of its attraction is evident in the sheer diversity of stone types in the debitage that surrounds the Tower. While there is a small local source of quartzite suitable for coarser tools, finer materials had to be brought to the site. Obsidian was the most exotic material, and all the fragments recovered were from several frequently used locations in western Wyoming and eastern Idaho except one, which came from an unusual source – the Mineral Mountains of southern Utah, a distance of almost 900 miles (Molyneaux 2002).

The earliest evidence of practical use was for reckoning. As the Tower rises vertically from its base more than 1200 feet, and has a collar of talus that extends part-way up the sides, it is possible to climb quite high without any technical equipment and look up and down the river valley and over the surrounding plains. There are much higher prospects in the region, but the Tower has one significant advantage; it combines this significant height with a very small footprint. As it is less than 1000 feet in diameter in any direction and only about a mile around, the climb is a matter of minutes from a camp below. For big game hunters on foot, speed would have been crucial. Bison often move quickly, and so a herd might appear only briefly, before disappearing behind hills or meanders in the valley or leaving altogether.

The earliest evidence of occupation at the Tower seems to relate to such reconnaissance. In a small niche high on the talus slope, several flakes of quartzite, chert and chalcedony, and the base of a large projectile point, were caught up in the roots of a burnt, toppled Ponderosa pine. The projectile had a precisely flaked rectangular base characteristic of an Alberta point, a regional style that appears between 11,000 and 9000 years ago. From this spot, a hunter may have spent some time watching for game along the river as it meanders north into the rugged hills at the edge of the Bearlodge Mountains.

The patches of soil and forest cover scattered among gaps in the talus around the formation also provided substantial shelter for more prolonged occupation, especially on the east side, where the mass of the rock protected occupants completely from prevailing winds and weather. Historic Sioux narratives mention winter camping here in the mid-19th century. It was "fine winter country … with all kinds of animals" where "the cold winds were kept out by the hills" (Oscar One Bull, Hunkpapa, 1934 (translated by Luke Eagle Man) and recorded by Stone n.d., 52). White Bull, a chief of the Minneconju, recalled camping there many times during his youth.

> Sometimes, years ago, we would go to Bears Tipi and stay all winter, that is how the arrows and scraping knives came to be found there. When I was two years old [ca. 1852] I spent the winter there with the Minneconjui, Itazipco and Uncpapa bands, these bands all speak the same language. They hunted antelope, buffalo and deer, there was also black bears and grey wolves around there. When I was fourteen years old we wintered at this place and again when I was eighteen years old. We wintered in different places around this hill each time.
>
> (Dick Stone n.d., 53–53A)

Along with the evidence of occupation, it appears from the archaeological record that the Tower affordances changed with changes in technology, just as they did at Samanakuta. Here, the evidence is in the absence of cultural material over two periods of time, when it seems that the Tower saw little practical use.

The first hiatus corresponds to a change in climate at the end of the Wisconsinan glaciation. About 8,500 BP, the region became warmer and drier, and this Hypsithermal interval persisted for approximately 3000 years, until about 5500 BP. To the extent that water and game were available there would have been settlement here, but streams such as the Belle Fourche would have dried up more frequently and flooded only during torrential rain or snowmelt. The nearby Black Hills would have had a much more favourable environment at this time, as its springs were able to sustain the ecology through the drought. When cooler and wetter conditions returned, the biota increased in diversity and rich grasslands supported increasing numbers of bison and other grazing animals (Frison 1991). The Keyhole Reservoir sites, less than 25 miles upstream along the Belle Fourche drainage, attest to the intensity of occupation during this period (Kornfeld, Frison and Larson 1995). Movement along the Belle Fourche and by the Tower would have been considerable, and projectile point types consistent with this later period are well represented.

The second gap in material evidence occurs during the Protohistoric and early Historic periods, marked elsewhere in the Plains by the appearance of trade goods and traders in the 18th century. Despite extensive searching, the survey around the Tower found no ceramics, metals or glass dating before the 20th century (Molyneaux et al. 2000). This decline in the practical function of the Tower probably related to the arrival of the horse in the Plains. For a society on foot, the Tower was essential as a landmark, an overlook and a shelter, but on horseback, people took hours to achieve what took days on foot.

As with Samanakuta, changes in technology had no effect on the Tower's significance as a spiritual landmark. Yet, even this relationship began to change when fur traders, U.S. military forces and settlers began to move in greater numbers into the interior.

In the western Great Plains, the vastness of the territory, the distance from the coasts, the generally dry climate provided a brief and small measure of protection. Tribes had successfully worked out their adaptation to the grasslands over more than 10,000 years and lived effectively by hunting and gathering and farming. Even as fur traders created a new economy along the river valleys, the Lakota people in particular were able to develop a strong position in the region and control the territory around the Black Hills (Hämälainen 2019). By the mid-19th century, however, they could not overcome the technological imbalance with the incoming culture and the sheer masses of migrants. All the Great Plains tribes resisted this invasion, through debilitating wars and explicit efforts to pacify and assimilate them, but by the end of the 19th century they were all forced onto reservations, much smaller and in most cases distant from their traditional lands.

This created a third aspect in the Tower's cultural life – when tribal people were prevented from living there. By the late 1870s, ranchers began to drive cattle up from Texas and settle in the eastern Wyoming grasslands (Driskill 1955), and in 1879 the town of Sundance was established about 27 miles from the Tower. With the institution of the National Monument in 1906, the Tower got a measure of physical protection, but the accompanying tourist developments began to exploit the same areas that were traditionally inhabited – around its shoulders – and this disturbed the ancient camps and pathways already there.

For indigenous people this was also a cultural barrier. From the beginning of the reservation period, they were denied the fulness of their historical relationships to the land of their ancestors, and this was only magnified when the Tower became part of the national interest. Ironically, its protection forced Tribes who had worshipped there for centuries to work within boundaries imposed by the dominant culture and, reluctantly, objectify their beliefs and traditions in order to speak out and resist.

Despite these depredations, indigenous people continue to assert their cultural independence. Whether or not a Tribe physically occupies its original homeland, its people maintain social and spiritual connections. Important places, such as the Tower, have therefore become icons that simultaneously represent traditional ways of life and the lands that each Tribe was once free to occupy, and that they continue to claim as their own.

Looking back to the first sighting of the Tower, the narratives and artifacts in the deep surroundings affirm that it was not a remote, spiritual place, but rather a focus in an integrated way of life that persisted over thousands of years. Its flanks provided commanding views of the river valley and its resources. The meadows, exposed among talus boulders and sheltered by the Tower, had level, well-drained habitation sites and work areas in all seasons. Springs and seeps attracted game and provided fresh water. Upland resources, the broad lower meadows, river terraces and the river valley were a short distance away. And spiritual resources were there in the history of its unusual shape, which was like nothing else on the earth. In all the evidence on the ground and in the oral histories, one thing seems clear about this great rock: long ago, it was a good place to live.

Monumental power

Monuments are not static places that humans choose to appropriate, construct or exploit to serve as theatres for ritual enactments or repositories for narrative. They are more like cynosures, with strong affordances that enable a wide range of possible material and social acts, depending on the environment, society and situation. While the contents of descriptions and the activities around them may vary, the common nature of these affordances gives monuments their drawing power.

The sheer energy in the salience of a monument makes it an effective ideological tool. As commanding features in their terrain, monuments are unrelated to the cultural matters that roil around them and pick away at their flanks. Their salience and physical presence provoke common action and their persistence increases the potential for patterns in interpretation. All it may take are a few physical details to create a consensus that crosses social and cultural boundaries. The Tower was for the Shoshone the stem of a gourd, for the Kiowa, a tree rock, and for many others, a rock that grew to save children from a rampaging bear – but all accounts converged on its remarkable furrows. And on Samanakuta, the hollow was surely the footprint of a sacred being – whether Saman, the Buddha or Adam.

This sensory directness is also how built monuments survive their makers. To the uninitiated, seeing the Lion's Mound may be enough to define the experience of visiting the battleground. To the initiated, the focus is the events of June 18, 1815, a battle endlessly re-engaged in debates about tactics, troop movements and the generals. Unfortunately, while this rich imaginative history casts for them an aura over the battlefield, these forensics never see the light of day. Only the Mound greets the eye. And if all memory of the Battle of Waterloo was lost, and the great lion and globe were scavenged for their metal, the mound would still engage, a Silbury Hill in the Belgian countryside.

The idea of drawing power therefore causes us to reconsider the sense of place, to allow reference to forces outside reverie, narrative and the human project. If a place is a locus of environmental interaction, it is easier to accept diverse interpretations because the truth of beliefs is simply not at stake. Set in their environments, the Tower equally affords the conclusion that it was a huge, ancient tree that some believe a giant bear tried to climb, while the pyramidal shape and great height of the Peak made it a suitable resting spot for a sacred being who had traveled through the heavens. The model is the world, resolved in the ordinary actions of daily life.

In the next chapter, we turn to visual art. A view focused on environmental interaction, at home among landmarks and habitation sites, may seem lost in pictures – yet, they are also environments that we are drawn to, and within which we act. The rest of this book explores this phenomenon, to see how the drawing power of huge monuments translates at the diminutive scale of paintings.

References

Abhayasingha, K.R. 2012. Scientific Explanation of Ira Sevaya. *Sunday Observer* (Sri Lanka), December 16, 2012.

Al-Masudi. [947] 1841. Meadows of Gold and Mines of Gems. In *El-Mas'udi's Historical Encyclopaedia Entitled Meadows of Gold and Mines of Gems.* Translated from the Arabic by Aloys Sprenger, M.D. Volume 1. London: Printed for the Oriental Translation Fund of Great Britain and Ireland; Sold by W. H. Allen and Co.

Al Tifaschi, Ahmad ibn Yusuf. [1253] 1998. *Arab Roots of Gemology: Al Tifaschi's Best Thoughts on the Best of Stones*, translated and with commentary by Samar Najm Abul Huda. Lanham, Maryland: The Scarecrow Press.

Bailey, John. 1863. An Account of the Wild Tribes of the Veddahs of Ceylon: their Habits, Customs, and Superstitions. *Transactions of the Ethnological Society of London* 2: 278–320.

Berkwitz, Stephen C. 2017. *Sinhala sandēśa poetry in a cosmopolitan context.* In *Sri Lanka at the Crossroads of History*, edited by Zoltán Biedermann and Alan Strathern, 94–112. UCL Press, London.

Blackiston, Major John. 1835. *Twenty Years in Retirement. In Two Volumes.* Volume 1. London: James Cochrane and Co.

Broun, A.F. 1900. On the Forest and Waste Lands of Ceylon. In *A Handbook to the Flora of Ceylon. Containing Descriptions of all the Species of Flowering Plants Indigenous to the Island, and Notes on their History, Distribution, and Uses. Part V. Eriocauloneae – Gramineae.* Edited by Henry Trimen and J.D. Hooker, 355–363. London: Dulau & Co.

Cattanar and P. Pandian. 1989. *Cattanar's Manimekalai.* Madras: South India Saiva Siddhanta Works.

Coomaraswamy, A.K. 1928. *Yaksas.* Smithsonian Miscellaneous Collections, vol. 80, Nr. 4. Washington: The Smithsonian Museum.

Davis, J.E. 2013. Plains Indian Sign Language: The Legacy of Documentary Linguistics Fieldwork. In *Keeping Languages Alive: Documentation, Pedagogy, and Revitalization*, edited by M.C. Jones and S. Ogilvie, 69–82. Cambridge: Cambridge University Press.

Davy, John. 1818. A Description of Adam's Peak. In a Letter addressed to Sir Humphry Davy, F.R.S., LL.D. *Journal of Science and the Arts* V:25–30.

Davy, John. 1836. *Memoirs of the Life of Sir Humphry Davy.* In Two Volumes. Vol. I. London: Printed for Longman, Rees, Orme, Brown, Green, & Longman.

Deraniyagala, S.U. 2007. The Prehistory and Protohistory of Sri Lanka. In *The Art and Archaeology of Sri Lanka*, edited by L. Prematilleke, S. Bandaranayake, S.U. Deraniyagala and R. Silva, 1–99. Colombo: Central Cultural Fund.

Dewey, John. 1929. *Experience and Nature.* London: George Allen & Unwin.

Dorsey, George A., and Alfred L. Kroeber. 1903. *Traditions of the Arapaho.* Chicago: Field Columbian Museum, Publication 81. Anthropological Series, Vol. 5.

Driskill, Jess. 1955. *Ms.* Letter in Devils Tower Archives about driving cattle up the Texas Trail. National Park Service, Devils Tower National Monument, Archives RF-223.

Faris, Nabih Amin. 1952. *The Book of Idols: Being a Translation from the Arabic of the Kitab Al-Asnam, by Hisham Ibn-Al-Kalbi.* Translated with Introduction and Notes. Princeton: Princeton University Press.

Frison, George C. 1991. *Prehistoric Hunters on the High Plains.* Second Edition. San Diego: Academic Press.

Geiger, Wilhelm. 1912. *The Mahavamsa or the Great Chronicle of Ceylon.* Oxford: Oxford University Press.

Geiger, Wilhelm. 1929. *Culavamsa; being the more recent part of the Mahavamsa. Translated from the German by C. Mabel Rickmers.* London: Pali Text Society.

Goodman, Ronald. 1992. *Lakota Star Knowledge: Studies in Lakota Stellar Theology.* Rosebud: Sinte Gleska University.

Gunnerson, Dolores A., and Gunnerson, James H. 1971. Apachean Culture: A Study in Unity and Diversity. In *Apachean Culture History and Ethnology*, edited by Keith H. Basso and Morris E. Opler, 7–27. Tucson: University of Arizona Press.

Hallowell, A. Irving. 1926. Bear Ceremonialism in the Northern Hemisphere. *American Anthropologist*1926:2–175.

Hämälainen, Pekka. 2019. *Lakota America: A New History of Indigenous Power.* New Haven: Yale University Press.

Herrmann, Albert. 1935. *Die älteste türkische Weltkarte (1076 n.* Chr.). Imago Mundi 1:21–28.

Hirth, F. 1896. Chao Ju-kua's Ethnography: Table of Contents and Extracts regarding Ceylon and India, and Some Articles of Trade. *Journal of the Royal Asiatic Society* 28:477–507.

Hirth, Friedrich, and W.W. Rockhill. 1911. *Chau Ju-Kua: His Work on the Chinese and Arab Trade in the Twelfth and Thirteenth Centuries, entitled Chu-fan-chi.* Translated from the Chinese and Annotated. St. Petersburg: Printing Office of the Imperial Academy of Sciences.

Holt, John Clifford. 1996. *The Religious World of Kirti Sri. Buddhism, Art, and Politics in Late Medieval Sri Lanka.* Oxford: Oxford University Press.

Hornaday, William T. 1889. *The Extermination of the American Bison.* Washington: Government Printing Office.

Hosten, S.J. 1936. *Antiquities from San Thome and Mylapore.* San Thome: The Diocese of Mylapore.

Hugo, Victor. 1887. *Les Miserables, Part Second. Cosette.* Authorized translation by Sir Lascelles Wraxall. Boston: Little, Brown and Company.

Kaplony, Andreas. 2008. Comparing Al-Kashghari 's Map to his Text: On the Visual Language, Purpose, and Transmission of Arabic-Islamic Maps. In *The Journey of Maps and Images on the Silk Road*, edited by Philippe Forêt and Andreas Kaplony, 137–154. Leiden: Brill.

Kennedy, Kenneth A.R., and Siran U. Deraniyagala. 1989. Fossil Remains of 28,000-Year-Old Hominids from Sri Lanka. *Current Anthropology* 30:394–399.

Kinnard, Jacob N. 2000. The Polyvalent Pādas of Viṣṇu and the Buddha. *History of Religions* 40:32–57.

Knox, Robert. 1681. *An Historical Relation of the Island Ceylon in the East Indies Together with an Account of the Detaining in Captivity the Author and Divers Other Englishmen Now Living There, and of the Author's Miraculous Escape.* London: Printed by Richard Chiswell, Printer to the Royal Society, at the Rose and Crown in St. Paul's Churchyard.

Kornfeld, Marcel, George C. Frison and Mary Lou Larson. 1995. *Keyhole Reservoir Archaeology: Glimpses of the Past from Northeast Wyoming.* Occasional Papers on Wyoming Archaeology No. 5. Laramie: Office of the Wyoming State Archaeologist.

Lee, Reverend Samuel. 1829. *The Travels of Ibn Batuta.* London: Printed for the Oriental Translation Committee and Sold by J. Murray, Albemarle Street; Parbury, Allen, & Co, Leadenhall Street and Howell & Stewart, Holborn, London.

Legge, James. 1886. *Fa-Hien's Record of Buddhistic Kingdoms, Being an Account by. the Chinese Monk Fa-Hien of his Travels in India and Ceylon (A.D. 399–414). in Search of the Buddhist Books of Discipline.* Oxford: Clarendon Press.

Lin, Derek. 2006. *Tao Te Ching. Annotated & Explained.* Woodstock: Skylight Path Publishing.

Lipman, Jonathan N. 1997. *Familiar Strangers: A History of Muslims in Northwest China.* Seattle: University of Washington Press.

Lower Brule Sioux Tribe. 2017. *Map of Lakota Place Names within the Great Sioux Nation Lands of the 1868 Fort Laramie Treaty.* Lower Brule: Lower Brule Sioux Tribe.

Mackintosh-Smith, Tim. 2017. *Two Arab Travel Books.* New York: New York University Press.

Marx, Karl. 1906. *Capital. A Critique of Political Economy.* New York: Modern Library.

Mattson, David J., and Troy Merrill 2002Extirpations of Grizzly Bears in the Contiguous United States, 1850–2000. *Conservation Biology* 16:1123–1136.

McCrindle, J.W. 1897. *The Christian Topography of Cosmas, an Egyptian Monk.* London: Hakluyt Society.

Meadows, William C. 2008. *Kiowa Ethnogeography.* Austin: University of Texas Press.

Meadows, William C. 2016. New Data on Kiowa Protohistoric Origins. *Ethnohistory* 63:541–570.

Molyneaux, Brian L. 2002. Exploring the Landscapes of Long-distance Exchange: Evidence from Obsidian Cliffs and Devils Tower, Wyoming. In *Geochemical Evidence for Long-Distance Exchange*, edited by M. Glasscock, 133–151. Westport: Bergin & Garvey.

Molyneaux, Brian. 2006. Topographic Scale as Ideological and Practical Affordance: The Case of Devils Tower. In *Confronting Scale in Archaeology: Issues of Theory and Practice*, edited by Gary Lock and Brian L. Molyneaux, 67–76. New York: Springer.

Molyneaux, Brian L., Nancy J. Hodgson and Rachel M. Hinton. 2000. *The Archaeological Survey and National Register Evaluation of Devils Tower National Monument*, Crook County, Wyoming 1997–1998. National Park Service Contract No. 1443-CX-1230–1297–00l. Vermillion: Archaeology Laboratory University of South Dakota.

Mooney, James. 1898. *Calendar History of the Kiowa Indians.* Seventeenth Annual Report of the Bureau of American Ethnology to the Secretary of the Smithsonian Institution, 1895–1896, 129–444. Washington: Government Printing Office.

Morrison, Kathleen D. In Press. Water in South India and Sri Lanka: Agriculture, Irrigation, Politics, and Purity. In *History of Water and Civilization, Volume VII, Water and Humanity: an Historical Overview*, edited by Y. Yasuda and V. Scarborough, UNESCO.

Mumford, Lewis. 1934. *Technics and Civilization.* London: Routledge and Kegan Paul.

Oldenberg, Hermann. 1879. *The Dipavamsa: an ancient Buddhist historical record.* London: Williams and Norgate.

Paranavitana, S. 1958. *The God of Adam's Peak.* Ascona: Artibus-Asiae Publishers.

Pasztor, Emilia. 2015. Nebra Disk. In *Handbook of Archaeoastronomy and Ethnoastronomy*, edited by Clive L. N.Ruggles, 1349–1356. New York: Springer.

Patterson, Jessica Lee. 2012. Ostentation and Invisibility: The Phra Phutthabat and Royal Pilgrimage in Late Ayutthaya. *Artibus Asiae* 72:159–192.

Percival, Robert. 1803. *An Account of the Island of Ceylon.* London: C. and R. Baldwin.

Pieris, P.E. 1920. *Ceylon and the Portuguese 1505–1658*. Tellippalai: American Ceylon Mission Press.

Poe, Edgar Allan. 1841. Descent into the Maelstrom. *Graham's Lady's and Gentleman's Magazine* 17 (May):235–241.

Poolsuwan, Samerchai, and Patcharee Lertrit. 2013. Mitochondrial DNA history of Sri Lankan ethnic people: their relations within the island and with the Indian subcontinental populations. *Journal of Human Genetics*, 1–9.

Pridham, Charles. 1849. *An Historical, Political and Statistical Account of Ceylon and its Dependencies*. Volume II. London: T & W Boone.

Raghavan, M.D. 1964. *India in Ceylonese History, Society and Culture*. Bombay: Indian Council for Cultural Relations, Asia Publishing House, Delhi Press.

Ranaweera, Lanka, Supannee Kaewsutthi, Aung Win Tun, Hathaichanoke Boonyarit and W.F. Raynolds 1868. *Report on the Exploration of the Yellowstone River, By Bvt. Brig. Gen. W. F. Raynolds. Communicated by The Secretary of War, In Compliance with a Resolution Of Senate, February 18, 1866*. Washington: Government Printing Office.

Riggs, Stephen J. 1869. *Tah-koo; or, The Gospel among the Dakotas*. Boston: Congregational Publishing Society, Congregational House.

Roberts, Patrick, Nicole Boivin and Michael Petraglia. 2015. The Sri Lankan "Microlithic" Tradition c. 38,000 to 3,000 Years Ago: Tropical Technologies and Adaptations of Homo sapiens at the Southern Edge of Asia. *Journal of World Prehistory* 28 (2):69–112.

Roberts, Patrick, Nimal Perera, Oshan Wedage, Siran Deraniyagala, Jude Perera, Saman Eregama, Andrew Gledhill, Michael D. Petraglia and Julia A. Lee-Thorp. 2015. Direct Evidence for Human Reliance on Rainforest Resources in Late Pleistocene Sri Lanka. *Science* 347:1246–1249.

Roberts, Patrick, and Michael Petraglia. 2015. Pleistocene Rainforests: Barriers or Attractive Environments for Early Human Foragers? *World Archaeology* 47:718–739.

Ross, E. Denison, and Eileen Power (eds). 1929. *Ibn Battuta Travels in Asia and Africa 1325–1354*. Translated and selected by H.A.R. Gibb. London: Routledge & Kegan Paul Ltd.

Sastri, K.A. Nilakanta. 1955. *The Colas*. Second Edition, revised. Madras: G.S. Press.

Scott, Hugh L. 1920. Letter, to the President of the Historical Society of Wyoming, May 10, 1920. In *First Biennial Report of the State Historian of the State of Wyoming for the Period Ending September 30, 1920 with Wyoming Historical Collections*. Laramie: Laramie Printing Company.

Seligmann, C.G., and Brenda Z. Seligmann. 1911. *The Veddas*. Cambridge: Cambridge University Press.

Shimkin, D.B. 1942. Dynamics of Recent Wind River Shoshone History. *American Anthropologist* 44:451–462.

Shimkin, D.B. 1947a. *Wind River Shoshone Ethnogeography*. Anthropological Records 5(4). Berkeley: University of California Press.

Shimkin, D.B. 1947b. *Childhood and Development Among the Wind River Shoshone*. Anthropological Records, Volume 5(5). Berkeley: University of California Press.

Silberman, Neil A. 2007. Reshaping Waterloo: History, Archaeology, and the European Heritage Industry. *Archaeology*, Volume 60, Number 11. http://scholarworks.umass.edu/efsp_pub_articles/11.

Sims, M.E. 2007. *Comparison of Black Bear Paws to Human Hands and Feet*. Identification Guides for Wildlife Law Enforcement No. 11. Ashland: USFWS, National Fish and Wildlife Forensics Laboratory.

Skeen, William. 1870. *Adam's Peak. Legendary, Traditional, and Historic Notices of the Samanala and Sri-Pada, With a Descriptive Account of the Pilgrims' Route from Colombo to the Sacred Foot-Print*. Colombo: W.L.H. Skeen & Co.

Stone, Dick. n.d. Devils Tower National Monument Library Archives. Dick Stone Manuscripts: American Indian Oral Histories and Local Histories. 1930s. *Vertical File History* #32.

Strong, John. 2004. *Relics of the Buddha*. Princeton: Princeton University Press.

Sundstrom, Linea. 1996. Mirror of Heaven: Cross-Cultural Transference of the Sacred Geography of the Black Hills. *World Archaeology* 28:177–189.

Tankersley, Kenneth B., Kevin O. Tankersley, Nelson R. Shaffer, Marc D. Hess, John S. Benz, F. Rudolf Turner, Michael D. Stafford, George M. Zeimens and George C. Frison. 1995. They have a Rock that Bleeds: Sunrise Red Ochre and its Early Paleoindian Occurrence at the Hell Gap Site, Wyoming. *Plains Anthropologist* 40:185–194.

Taylor, Philip. 1995. Jumping Spiders Alternate Turns in the Absence of Visual Cues. *International Journal of Comparative Psychology* 8:69–76.

Thapar, Romila. 1971. The Image of the Barbarian in Early India. *Comparative Studies in Society and History* 13:408–436.

Thwaites, R.G., ed. 1905. *Original Journals of the Lewis and Clark Expedition, 1804–1806*, Volume 4. New York: Dodd, Mead & Company.

Tuan, Yi-Fu. 1979. Space and Place: Humanistic Perspective. In *Philosophy in Geography*, edited by S. Gale and Gunnar Olsson, 387–427. Boston: D. Reidel.

Tweet, Justin. 2019. *Paleontology of Devils Tower National Monument*. Park Paleontology News 11, No. 2. https://www.nps.gov/articles/park-paleo-fall-2019-devils-tower.htm.

Uffindell, Andrew, and Michael Corum. 1966. *On the Fields of Glory*. Mechanicsburg: Stackpole Books.

Waterman, T.T. 1914. The Explanatory Element in the Folk-Tales of the North-American Indians. *Journal of American Folklore* 27:1–54.

Winnecke, Friedrich. 1878. On the Visibility of Stars in the Pleiades to the Naked Eye. *Monthly Notices of the Royal Astronomical Society* 39:146–148.

Woodhouse, L.G.O. 1950. *The Butterfly Fauna of Ceylon*. Second Abridged Edition. Colombo: Ceylon Government Press.

4 The grounds of art

Rock art

Rock art by its nature has elusive meanings and functions, but it is firmly grounded.

On the great mass of rocks that make up Grey Horn Butte, there are two tiny rock painting sites. The images are very small, heavily weathered and difficult to see. One is on the back wall of a small niche on the upper slope of the Tower, a jumble of red ochre lines in brown and purple shades that covers a small patch of unusually smooth rock. The surface is heavily weathered and rapidly exfoliating, and the lines are faint. The other painting survives as a narrow, reddish wash on the lip of a small, narrow vertical crevice near the bottom of the tower. It is so eroded that an exfoliated fragment had to be analyzed to confirm that it was red ochre and not a natural stain (Molyneaux et al. 2000).

Petroglyphs and rock paintings are scattered across the Black Hills region (Sundstrom 2004). What is unusual here is that the vast rock surfaces of the Tower, its talus fields and surrounding bands of sandstone have been available for painting and carving over at least 10,000 years of occupation, yet only these two small locations are known. As the Tower has provoked curiosity and reverence through recorded history, and rock art in general is interpreted as a form of spiritual expression, this is hard to explain.

The ordinary solution is to apply information from other sites and ideas from other places – we must say something. The search for meaning and function may even draw on current traditions, without any evidence as to the age of the images. There is one context, however, that is still available to us: the grounds within which the painter worked. This makes sense. The archaeologist who surveyed here also climbed up the slope of the Tower mound, walked around the base of the rock, climbed up to the crevice and the niche and gazed at the paintings and their surroundings.

In these shared circumstances, some possibilities about the making of the art emerge. The painters chose these locations with great care, a difficult task on such a steep slope, as each has unusual features in this already extraordinary setting. In one place, it is the patch of rock and the crevice. The

DOI: 10.4324/9781351053105-4

whole formation is riven with huge vertical joints between the columns of coarse porphyry, but this small patch is smooth, and it is above a narrow foothold in the almost vertical slope. The other site is a smooth patch of rock just above the surface of a niche in the slope large enough to accommodate a small group of people. Here, there is also an extraordinary view – one almost as striking as the sight of the Tower from a distance. Looking to the west, the rounded top of one of the nearby volcanic intrusions, known as the Missouri Buttes, dominates the horizon, but it is oddly cut off by an intervening ridge, making it appear like a great dome. This lends even more strangeness to an already strange place.

Within these settings, the acts of painting appear brief and straightforward. At the crevice, the painter daubed or smeared pigment around and inside the opening, possibly using it as part of an image. At the niche, the painting was vigorous, but it seemed to be focused on the act, consisting of arcs made by sweeping painted fingers down the rock face and, possibly, some handprints.

Considering the effort it takes to climb the slope and find these suitable places, the acts were probably not random – and this makes them even more curious. It is as if the choice to paint had never occurred to any other visitor in the thousands of years of intense activity at the Tower. If we forget about meaning in the art and look at the decisions the painters had to make here, however, some possibilities emerge.

Like Samanakuta, the Tower has an overwhelming presence and, although much smaller, its unique form, which gives it an unsettling, alien appearance, has inspired as many dramatic accounts of its origins. In such a climate, the daubing of pigment seems inconsequential – unless the painters were outsiders, without knowledge of local beliefs and practices, who treated this powerful landmark as they would in their own homelands. The shades of ochre, the finger draggings and the attention to the crevice recall Algonkian rock paintings of the kind scattered along the western margins of the Canadian Shield near the Lake of the Woods (e.g., Dewdney and Kidd 1967). Similar techniques might very well be found nearer to home in the plains and intermontane regions west of the Black Hills.

Long-distance travel in search of food and other resources was an ordinary part of life for hunting and gathering peoples, and oral traditions, archaeology and ethnohistory all record great treks across the continent. Many of these in the central part of the continent might have taken routes past the Tower, looming over the river in the mountainous bottleneck that divides the high plains of Wyoming from the rest of the Great Plains.

The question of why the painters painted there and where they came from is, of course, impossible to answer – there is too little material evidence to go on – but this modest effort to approach the issue at the Tower is intended to make a point essential to what follows. A work of art of any kind is not simply a product, existing in a context of relations to other products and representations. It is the outcome of an act in a live environment of direct relations with its surroundings, a single instance in an ecology of art.

In this approach, the apparent product is the result of a cascade of decisions, some of them substantive enough that we may be able to identify them in the persistent surrounding environment. For rock art at the Tower, each of these images involved journey, preparation and creation – that is, the situation of production.

A situational approach, with its interest in human–environment interaction and production, falls on the periphery of traditional studies of art. Within a totalizing view of culture, what art seems to encompass is assumed to be fundamentally human, ideologically configured and closed to outside worlds. To make a situational analysis possible, it is therefore necessary to forget what art is and, in its place, consider what it might be in an open field of action possibilities that extends outside social mediation and meaning. In the absence of meaning, the maker and all spectators, past and present, still share fundamental senses and kinetics in a persistent natural environment – a nature that includes the manifest behaviours celebrated in culture and society. As this integrated world of culture and nature exists outside a history of art, the study of picture-making needs to find new grounds. Here, it is display.

The art of the senses

The propensity for display in humans extends deep into the biological past, in the evolution of sensory systems and the ecologies of their use (Dusenbery 2002). Many living things have salient features that afford or constrain interactions with others, within or outside their own kind, and these attractants and displays provide some selective advantage. In most, the displays are passive. Plants may have brightly coloured flowers and fragrant perfumes, but they depend on insects for pollination, while equally colourful male birds may spend their days primping and singing to attract females, but they only succeed if females choose to respond to their displays. Only in hominins is display externalized and used consciously as a tool to manipulate attention and enhance communication.

Thinking of art in this way transforms its study in archaeology. If ancient human ancestors did not have to wait for speciation or the evolution of the modern mind to display with strategic intent, it makes no sense to try to push the origins of art into the past. Display must come forward. Once display replaces art as an organizing principle, the study of visual things is released from a dependence on representations and other forms of ideology to account for its presence and effects. In keeping with visual ecology, which deals with the ways that animals pick up and use visual information, a picture becomes a physical environment with a distribution of salient features intended to attract and direct attention for purposes that the maker intends. What makes a picture accessible in the absence of meaning is that some of these features are commensurate across human populations.

Where the difference with other animals appears is, by this definition, more in the complexity of this intent. One might therefore assume that the behaviours associated with strategic displays and, ultimately, art, are part of the rise of culture. However, some other animals create displays external to the body that show both the awareness of surroundings and the recognition of salient properties in things. In effect, they modify salience as an active strategy.

Many animals have evolved with camouflage, but at least one species actively camouflages itself – the suitably named decorator crab (family Majidae). As it moves through its marine habitat it fixes local vegetation to hooks on its shell to blend in with its surroundings – an effective way to avoid being eaten (Ruxton & Stevens 2015). The crabs somehow assess their circumstances and surrounding environment in this task, as their choice and extent of decoration varies as the vegetation does and, in some, changes in the presence of predators.

It is the bowerbird (Ptilonorhynchidae), however, a native of the forests of Australia and New Guinea, that has become the main foil for discussions about the boundaries between humans and other animals, including ideas about the nature of art (Endler 2012). The males in some bowerbird species do not depend on their colouring or calls to attract females. Male great bowerbirds (Ptilonorhynchus nuchalis), for example, weave bowers out of sticks on the ground and decorate them with a wide range of curious and colourful objects, from dung to berries and flowers, to plastic. Each male spends a great deal of time choosing a location, managing its bower, adding, changing, rearranging, mending and defending (Schaedelin and Taborsky 2009) – like a landscape gardener or interior decorator.

The capacity for choice is not the novelty here. The bowerbird's intent may go further. One species, the satin bowerbird (Ptilonorhynchus violaceus), even decorates the bower itself, by masticating plant matter into a paste and smearing it on the interior (e.g., Bravery, Nicholls, et al. 2006). Scent might be the motivation rather than colour, but Endler el al. (2005) show that some species will select colours outside the range of their plumage patterns to increase the contrast to the background. What is significant about ornaments in display is that they are "an innovation that increases signal efficacy" (2005, 1815). The birds are making strategic choices with salience as a form of signalling strategy (Schaedelin and Taborsky 2009).

The acts of these creatures relate entirely to the worlds they inhabit, which are in many ways different than our own, but they are striking and familiar to us because we too evaluate, plan, choose and draw or deflect attention in our own lives. It may therefore be appropriate to take a pragmatic approach and look for connections rather than divisions, as indigenous societies do, and bring humans back into the greater community of life.

Treating art as display may be especially helpful in the study of prehistoric art, as meaning is absent. Indeed, as display behaviours are grounded outside the human project, the idea that art originated in the Paleolithic needs to be reconsidered. If image-making emerged within adaptive processes associated with attention, it cannot be the spark of human culture and so automatically attributed to cognitive evolution, the rise of spirituality or any other phenomenon with immanent cultural meanings and functions.

What replaces this cultural dynamic is sensory ecology, the way humans, like other species, use sensory information in daily life. We will assume for practical purposes that the urges and imperatives of display are the workings of sensory ecology – a dynamic that may ground and enhance the search for meaning in art.

Display in the Paleolithic

Ancient evidence for display in hominins appears first in portable items (ornaments, carvings and decorated objects) and rock art. Most rock art has a ready-made quality that suggests it was the work of ordinary people, created in exceptional moments in the course of daily life, but some examples, especially in caves in western Europe, show a much more specialized skill. It is not a primeval or primitive art; the practice continues in urban terrains in graffiti and other surface arts.

How visual display reached this point through early hominin evolution is unknown, and analogies to other primates yield little. Our nearest relatives, chimpanzees, may be trained to wield paint brushes, but they seem disinterested in visual display in the ways we ordinarily recognize (Van Schaik 2016). An unusual outlier in visual display is the Makapansgat Pebble. This odd-looking stone was found in dated sediments in an Australopithecine habitation so it seems that one of them picked up the stone from a gravel bed in the area and carried it into the shelter for some purpose. Why this creature did so may relate to natural markings that, to modern viewers, resembles a face. Archaeologists did consider the problem of conceptual bias (e.g., Dart 1974) but, clearly, the stone exhibited some visible property that drew the creature's attention and caused the urge to possess it.

Australopithecines were butchering meat and making and using tools at this time, at least 3.3 million years ago (McPherron et al. 2010), and so gathering and curating objects is not surprising. What is interesting about this example is how something in its appearance motivated two closely related creatures in similar ways. One, the Australopithecine, picked it out of the gravel and took it back to the shelter, and the other, an archaeologist, dug it out of the shelter and looked it over, with equal curiosity. Is this commensurate act too much of a leap in space, time, culture and species? It is much less so than between human and bowerbird, picking through forests for an appropriate treasure, or the crab and its self-adornment.

What seems to change in hominins is not something preternatural, whether from some sudden, induced change or a mysterious urge to culture; it is the shift of visual display from a passive presentment to an active instrument that communicates intentions. This is communication in McLuhan's (1964) practical sense. It involves a carrier, which is a medium that is material and persistent in the environment, and messages, the local, ephemeral transmissions of information intended to draw attention and provoke some form of physical or ideological action. Thus, while the bowerbird actively assembles a display site and passively awaits reactions (working with the medium), some humans and Neanderthals during the Pleistocene began to promote their external displays directly through interaction with others (working with the message). This is especially so with ornamentation. What sets beads and other external adornments apart is that they are carried on the body and so literally project displays and intentions into social situations.

The emergence of active, strategic display appears first in coloured pigments. Barham states that by 270,000 years ago, "Central and East African hominids had incorporated color into their lives" (2002, 189). Direct evidence of display begins to appear within the past 150,000 years. Ornamentation, in the form of perforated marine shells, red and yellow colourants, and shell containers with complex pigment residues, has recently been discovered in the Cueva de los Aviones in Iberia (Hoffmann, Angelucci et al. 2018). U-series dating of the flowstone that caps the deposit gives these finds an age of about 115,000 to 120,000 years. Caves and shelters in Africa reveal a variety of decorated materials dating to the middle Pleistocene, such as Blombos cave, on the Southern Cape coastline east of Cape Town, South Africa, which has geometric decorations, engravings on ochre nodules, and shell beads in sediments dated at ca. 73,000 BP (Henshilwood et al. 2018); and Diepkloof Rock Shelter, in Western Cape, with geometric engravings on ostrich shells about 60,000 BP (Texier et al. 2010).

The change in display from an entirely passive trait to an active communication comes with the depiction of visible things. In another Iberian cave with Neanderthal associations, an individual placed a hand on the cave wall and sprayed red ochre pigment over it to create a negative handprint; the U-Th (uranium-thorium) date on carbonate crusts over the image indicates a minimum age of over 66,000 years (Hoffman, Standish et al. 2018). Perhaps this was evidence of self-consciousness, an awareness of the separateness of one's own body. What we know for certain, however, is that this individual used its body as a tool in a public display.

The crucial difference between body decoration and external depiction relates to the direction of attention. Whereas the decorated body says "look at me", the picture on a wall says "look at this" and so draws attention to a much larger and more complex world. From this point, humans and Neanderthals carry display forward into self-conscious strategies related to visual engagement, reinforcement or communication.

The earliest evidence of this novel form of display is a painting in Leang Tedongnge cave in Sulawesi, which has a minimum age of 45,500 years (Brumm et al. 2021; and see image online). It is a detailed image of a pig, painted in a reddish pigment. With the rise of parietal art in western Europe in mind, where Maidagan and Moro-Abadía (2012) noted that some of the earliest examples suggested a mature art form, the skill involved in this much earlier depiction suggests that this was not the painter's first work. What seems most remarkable, however, is that after 45,000 years of putative cognitive development and culture change we can still bear witness, identify the creature (Sulawesi warty pig (Sus celebensis)) and find its descendants, now domesticated, foraging in the same surroundings.

The picture must have had some meaning and significance to the painter and others, but this interpretive path dwindles outside the cave. Above the rump of the pig, however, there are two stencil handprints. These have not been dated, and so their relationship to the pig image is unknown. Yet, they have an interpretive advantage over the pig, because they show a clear intent on the part of the artist who put them there. In this way, they incorporate both self-display and the picturing of an external world, and this suggests a conscious strategy.

Such intent is abundantly clear in Leang Balangajia cave, where another pig, this one dated to about 32,000 years, has four superimposed hand stencils, three below on either side of the hindquarters, and one in the middle of the animal (Brumm et al. 2021; and see image online).

The imaginative possibilities are endless. Perhaps it was a way of marking, surrounding or claiming the more ancient image, or something in its body parts. But the relation is not. Through some urge or some reason, the painter arranged the prints with some care. The three on the edge intrude only slightly into the picture, while the other is carefully centered.

Seeing this picture through its display, in a material environment that combines the actions and perceptions of makers and viewers, cannot compete with the ideological implications of this extraordinary find in human history and ideas. However, it helps prepare the ground for future studies in ways that might help free future interpretation in Paleolithic art from conceptual habits developed over more than 2,000 years of intellectual humanism.

From the point of view of an art that developed in step with a cultural or symbolic mind and a mental language, there is no grounding, only patience, the long wait for time and cognitive change to transform Homo sapiens into modern humans. This allows for considerable research and interpretation, but the end result is distant from where it began – the pictures in the caves. If material things are only defined by their immateriality, then archaeology has not moved beyond its basic curiosity with artifacts.

A study of art as display cannot support the treatment of material things as carriers of culture, as if millions of years of adaptation and sensory and visual ecologies only mattered to other animals because humans received the gift of mind and imagination. Just as pinnacles and monuments get in the way of the unconstrained imagination, pictures cannot also be cultural isolates. The idea of display therefore helps resist the lure of representation, which draws images into the trap of the symbolic imagination.

Consider the simple rock painting of a moose (Alces alces) painted on a rock, somewhere in northern Canada. Obviously, a hand-sized, red ochre painting is not a moose, but as Gibson has observed, while a picture may not have any projective correspondence with the animal, the picture “supplies some of the information for what it depicts” (1978, 231). And indeed, in simple outline, it is easy to recognize, because of its distinctive antlers.

Still, there is a common, tacit assumption that the picture *represents* something else – a vague suggestion, a symbol or token, as the *Oxford English Dictionary* defines it (Simpson and Weiner 2009) – an idea. Indeed, why would the artist paint it, and what would be the point of interpretation if there was only description? On top of these assumptions there are others, that what the picture represents exists as an image or schema in the mind, and that it is not something conjured up by the artist, but is the product of a collective idea, somehow shared with other members of the painter's society and reprocessed over time.

This is a problem for understanding the thing that one looks at on the rock. If it is the center of attention, as an embodiment of distant meanings and functions, any earthly connections with its artist, community and surroundings dissolve. A symbolic moose never lived its life browsing in the muskeg of a northern forest, or served the needs of people who lived nearby. As for the picture on the rock, there is no need to explore it further. As a representation of something else, it is dead to the world in which it was created.

Lascaux

The absence of grounding in interpretation is especially clear in the study of the Upper Paleolithic imagery of western Europe. In Lascaux cave in southwestern France, the hundreds of engravings and rock paintings were created so long ago that there is no ecological or social continuity between the remote past and the present.

The setting provides an ideal rock art problematic. The paintings and carvings are in precisely configured interior spaces (except the floor), with limited relatable ground outside the entrance. The cave therefore forms the physical extents of its world, and so by default serves as its own model. In almost 80 years of intensive research, it seems that every nook and cranny has been explored, every subtlety of surface and topography mapped and considered in relation to the imagery, images reproduced in all extant media, pigments analyzed, residues of production found and studied, and cultural materials and other evidence of activity dug up from the floors.

When the Abbé Breuil saw the site in 1940, he and others assumed that the art related to hunt narratives or hunting magic (Bandi, et al. 1961), ideas based on the practices and speech of more recent hunting and gathering cultures, and assumptions about the nature of pictures and their function in cultural life. However, if a picture of a horse was at least superficially a horse (by far the most common species pictured in the cave), it became increasingly involved in matters outside its ordinary life. Most influential was the interest in the later 20th century in the activities of religious specialists in hunting and gathering, herding and horticultural societies who used animal objects and images and ecstatic methods in divination: shamans.

Shamans were first named and described in the accounts of European travelers to north-central Asia in the 17th century. When Brand (1698, 48–51) wrote an account of visit to a Tungus village in east Siberia, he described what was clearly a well-developed ritual tradition.

> If five or six of these Tonguese Families happen to live near one another (which is very rarely seen) they maintain betwixt them a Shaman, which signifies as much as a Sorcerer or Priest. At their Meetings this Priest appears in a Habit distinguish'd by many Figures of Beasts, such as Lions, Bears, Serpents, Adders, and such-like; which being all of Iron, weigh generally above Two hundred pounds weight. Then he beats a Drum in a very doleful manner, at which the standers-by break out into most dreadful Lamentations and Outcries, pretending that they see certain Spectres in the figures of Ravens and other strange Birds; and the Priest, as if strucken with an Epileptick-fit, falls down upon the ground …

Travelers and, eventually, ethnographers observed similar uses of animal imagery, performance and visions in other parts of the world, and shamanism became identified as the primeval form of spirituality. When Eliade (1964) attempted to clarify the concept of shamanism and differentiate it explicitly from other forms of religious practice, he used north Asian shamanism as its core, calling it shamanism's "most complete manifestation" (1964, 6). The intent was clear. Shamanism may have often been muddled by religious influence from other cultures, but it emerged from deep time as "a whole spiritual world, which, though differing from our own, is neither less consistent nor less interesting" (1964, xx).

This was an obvious boon for prehistoric art in general. Shamanic religiosity must have emerged fully in these societies, rather than as a primitive precursor to western religion, and at Lascaux the chambers and passageways were full of paintings and carvings of animals. This was a fortunate conjunction of isolates: the visionary from ordinary society, the art from its makers and the caves from the modern terrain around them. Shamanism provided what was missing, a fully developed social context that would bring the art into an intellectual world.

The alignment of the art with shamanic ritual intensified the already seductive visual power of the images, with their blunt styles and immense age. In chambers and passageways decorated on walls and ceilings with colourful animals – including the black outline of a huge bull that extends 17 feet along the passage – it is easy to imagine shamans in the throes of vision, lit by flickering torches and perhaps watched by others.

Predictably, research interest in the less colourful pictures and in engravings was more subdued. The dull and complicated palimpsests of engravings in Lascaux, numbering more than 1,000, remain relatively unstudied and uninterpreted. Salience and simplicity influence choice across all ideological boundaries.

The religious spectacles imagined in the cave do not complete its sanctification. The naming and descriptions of the spaces in Lascaux cave mimic the architecture of a Christian church, with a nave, apse and sanctuaries. The convergence of these two powerful spiritual ideologies extended to artifacts excavated from the floors. Among the many lamps discovered, Glory (1961) interpreted the most spectacular one as a "brûloir à fumigation" (1961, 182), an ancient version of a thurible or smudge pot.

Such analogies are inevitable. We are bound to interpret. Given the environmental and social isolation of the caves, however, and because of the certainty that the lives of hunters and gatherers were infused with spirituality, there was little rein on comparative evidence. Comprehensive ideas about underlying modes of belief quickly appeared – Campbell's idea of a singular myth that unites humanity behind its many variations (1949 and 1960), Eliade's primeval spirituality (1964) and, eventually, Fodor's concept of mental language (1975) and more recent studies of cognitive evolution and language (e.g., Mithen 1999), which seem somewhat closer to the groundwork that Franz Boas (1911) laid.

The ideological approach is clearly seductive, as it provides relief from the sparseness of the evidence and provides a discipline within which to organize interpretations. This is especially important in Lascaux, as speculation has ranged wildly, from dour functionalism to flights of fancy inspired by the aura of mystery and sanctity that seems to reside in the caves. Unfortunately, it is also limiting, as the appeal to representation draws attention away from its fundamental source – the unyielding paintings and engravings on rocks and cave walls.

The scene in the Shaft

These various and contrary flows come together in a small panel at the bottom of a fissure below the main passageways, commonly known as the Shaft, where a bison, a rhinoceros, a human-like stick figure, a bird on a stick and several linear drawings, all painted with a blackish pigment, are crowded together. There is no physical evidence that might confirm the contemporaneity of any two images on the rock, but seeing this group as a coherent scene seems almost irresistible (see discussion in Bahn 1995).

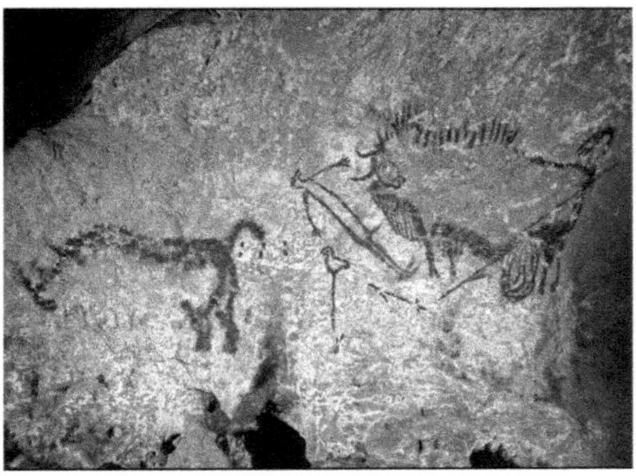

Figure 4.1 The Scene in the Shaft, Lascaux

Breuil, who first saw the images in 1940, described them in a typically straightforward way (Bandi et al. 1961, 36):

> … a badly wounded bison, its flank pierced by a lance, its horns lowered ready for the charge, and before it, sketched as always with a few strokes, the figure of a man is seen prostrate on the ground; in the foreground is a bird on a perch, and to the left a rhinoceros can be seen moving away. The whole composition is drawn in black, with the lines slightly blurred. It has been interpreted in many different ways; it is probably either a tragedy of the chase or a scene of a magic ritual character.

Soon after the discovery, however, Francoise Henry (1941) described the *scene*, and this narrative begins a vast literature on the subject.

> … a rhinoceros is depicted walking away from a scene of carnage for which he seems partly responsible, a man lies on his back, seemingly dead, and a wounded buffalo beside him is disemboweled like a "corrida" horse. The man's javelin and throw-stick lie on the ground, whilst a schematic bird on top of a pole seems to be a totemic emblem.

A little over a decade later, Breuil sent an angry letter, published in the July 1952 edition of the journal *Man*, complaining about "arbitrary fantasies", interpretations of the animal figures "as a symbolic representation of human beings of such and such totem", and concluding that "the magic of reproduction, of hunting and destruction … until further notice, suffices to explain everything we know about Quaternary art" (Breuil 1952, 111). By this time, however, studies of religion among indigenous people were beginning to converge around shamanism, and Joseph Campbell's interpretation of the Lascaux tableau in 1949 throws caution to the wind:

> Down there [in The Shaft] a large bison bull, eviscerated by a spear that has transfixed its anus and emerged through its sexual organ, stands before a prostrate man. The latter (the only crudely drawn figure, and the only human figure in the cave) is rapt in a shamanistic trance. He wears a bird mask; his phallus, erect, is pointing at the pierced bull; a throwing stick lies on the ground at his feet; and beside him stands a wand or staff, bearing on its tip the image of a bird. And then, behind this prostrate shaman, is a large rhinoceros, apparently defecating as it walks away.
>
> (Campbell 1960, 301)

All the readings of this group of figures illustrate the simple problem that archaeologists face at the abyss between tangible things and the meanings they once had. Interpretation, at least in institutionalized settings, responds to the ideas of the day, a revolving discourse of theories about culture, society and individuals. In the Shaft, debates have revolved between notions of the painter as a capable witness to an historical or, at least, representative event or situation, and as an unwitting proxy in a grand ideological tradition. Either way, the nature of discourse requires that the analysts leave the field of action and its traces of the painters behind. In all these ambitious schemes, the art has the psychological taint of the exceptional. There seems to be no concept of an ordinary painter with an ordinary life outside the cave.

A brief example among many illustrates how the interpretive conventions, the demands of discourse and the absence of restraint can influence what analysts see on these rocks. In 1988, Davenport and Jochim offered what they described as a new interpretation of the shamanic ritual they saw in the panel. They suggest that the close association of the wounded bison and the human figure "indicates a single artist with a story to tell" (1988, 560).

With the scene's integrity resolved, they join others who see birdlike features in the stick figure: the projection of the face in two lines as a beak; and the hands as the four toes of a bird, which they describe as "a deliberate and, indeed, sophisticated ploy of the artist to make the image more bird-like" (1988, 560). Moving down the figure, they treat the rest of the torso, arms and legs and erect phallus as human. With these details, they conclude that the creature is half man, half bird. Turning to the bird on a stick, they identify the species – a black grouse (Lyrurus tetrix) or capercaillie (Tetrao urogallus), both having a wide range across Europe and Asia.

From these results, they see the representation of a dramatic event. A shaman, mortally wounded, transforms into its grouse familiar through the mating dance that gallinaceous birds perform on their leks (dance grounds). The erection on the stick figure signifies (and registers) the desires of the dancing grouse.

The crux of this argument relates to the hands of the stick figure. They state that the four fingers represent the four toes of a bird, and they express wonder that anyone might even be skeptical:

> One is constrained to wonder what the reaction of the artist would have been if told that 14 or more millennia after his death, many authorities on his art would have so lost contact with the natural world as to be unaware of the significance of his putting four "fingers" on each hand.
> (1988, 560)

The problem on the rock face is that the four fingers in the figure are arranged not like a bird's toes but, rather, are stuck on the end of the stick arm in the same articulation as human fingers. Grouse and many other perching birds have three toes forward and a toe in back (anisodactyly). This is clearly visible in their footprints, which ordinarily appear as prominent three-toed tracks, with the fourth digit, facing back, often more lightly imprinted. These prints would have been as common a sight during the Pleistocene in the snow or along a muddy shoreline as they are today.

This suggests that the painter created an expedient image with just enough detail to identify it as a man. In the prevailing atmosphere of the time, however, this was difficult to argue. After all, as the Venus figurines and the lion man shows, sculptors of the time understood how to model the volume of human figures, and these figures are tacitly accepted as ritual objects. The man in the Shaft must therefore have been sketchy for some social or psychological reason. Thus, Davenport and Jochim carry on, supporting Kirchner's (1952) guess that the artist was reluctant to personalize the image because of the shaman's power, and they extend this supposed reticence to the sketch of the bird (1988, 561).

There are simpler solutions. Without the background of ideological associations, the human and bird on a stick are desultory sketches compared to the carefully crafted animals, and this might lead someone to wonder if the figures that took the most effort were the most important figures. This is not allowed, however, given the consuming interest of people in each other, the rarity of human images and the huge emotional and intellectual investments of interpreters in the concept of shamanism.

More than a decade later, David Lewis-Williams (2002) briefly mentioned this work. He agreed with their analysis of the bird-man, and he lauded their resistance to literal interpretation, but he states that there are "aspects of Upper Paleolithic art with which *they* are not familiar" (emphasis added). Lewis-Williams maintains that the bison is a spirit bison, and because its hoofs appear to be cloven, "it is 'floating' in spiritual space" (2002, 265). Notwithstanding that the only people familiar with Upper Paleolithic art are the people who made it, the issue is moot, because to get to this point in analysis has already involved a cascade of assumptions. Consider the bird on the stick. Other non-galliform birds resident in the region now and during the Pleistocene have virtually the same simple profile. The European golden plover (Pluvialis apricaria), for example, easily fits the sketch. It also breeds on arctic tundra and favours coastal and interior shorelines and damp inlands. It was also present in the region in the late Pleistocene: Mourer-Chauvire (1975) recovered this species in the Alps in excavations at Les Grottes Jean-Pierre, Savoie, in a layer dating to about 13,000 BCE. It, too, has been recorded elsewhere as a spirit helper (Jochelson 1905). But it does not dance.

When a coherent body of concepts representing a mode of being – Eliade's "whole spiritual world" – informs research from the beginning, there is little need to dwell on the local form and material context of individual pictures. As the existing narrative has no particular ground, all it needs is a resemblance, and speculation can proceed unabated. The concept of shamanism roused analysts from the dull rounds of empirical description and classification, and such speculation apparently made sense – but Davenport and Jochim would not attempt this approach in biology. Only with the poetic licence that prehistoric art history provides do artists cower, shamans transform or bison float.

In looking at the pictures again, an important issue is the integrity of the scene. Breuil noted that the rhinoceros was painted in two stages, first a sketch and then sprayed pigment (1950, 360). This animal was always an odd feature – thinking back to Campbell's idea of it fleeing the scene, defecating as it went. It turns out that its pigment is different from the rest (Chalmin et al. 2003, 1595) and this opens up a whole new world of possibilities. Perhaps one painter worked there in two episodes, changing paint pots or techniques, or there were two painters at the time, or the lumps of pigment came from different sources, or a second painter added the rhino to the scene at some later date, as it would not have run out of a scene that did not exist.

All the actual recorders of Paleolithic art faced extremely difficult circumstances and they are allowed the views from their direct experience. Had they the benefits of 3D scanning, scanning electron microscopes and direct dating technologies, they would have sorted and organized the imagery with much more precision, and their original ideas might well have held (cf. Bahn 1995). Only changes in the ideology of governing narratives and institutions can shift the search for meaning, however, as it always takes place outside the cave. Thus, after many generations in which underlying religious concepts and humanism have dominated interpretation in Paleolithic art, Lascaux has become its own narrative, an organic representation that emerged in earth and stone, was nurtured and sustained by discovery and archaeological research, until the search for meaning extended far from its grounds. Now, with the authority of almost 80 years of narrative accretions, the ideological walls seem impenetrable. The cave is closed and fragile, new research possibilities are limited and the discourse has little to feed on but itself.

If only the cave that generations of specialists have studied over the years could be accepted as the center of its world. This would at least preserve the singularity of the artists, perhaps leading to a greater understanding of their efforts and, in the absence of meaning, allow some space for the imagined life of ancient communities in the Vézère Valley, rather than the forests and steppes of northeastern Asia.

Display and community

The idea of display and its grounded interpretation is perhaps clearer outside the charged atmosphere of Pleistocene caves. In the open air, where parietal art becomes rock art, there are many places where descendants of ancestral painters and carvers continue to interpret their work.

Above Rock Creek, a tributary of the Yellowstone River in south-central Montana, there is a sandstone outcrop with hundreds of petroglyphs, the work of hunters and gatherers who lived and traveled along the waterways between the intermontane Big Horn Basin to the south and northern plains. This is Apsáalooke (Crow) traditional territory and some of the images depict men wearing their hair in the traditional Crow way. The style is well-known in the region. Lakota people, who live south of the Crow in present-day South Dakota, call one butte in their country *Apsaloka Paha* (Crow Hill), because its outline reminds them of this hairstyle (Lower Brule Sioux Tribe 2017). The style therefore asserts that this is Crow heritage on the rocks (Loendorf and Porsche 1985, 20). Many of the depictions are recent, as they include many horses and scenes of horse-raiding and victory dances (1985, 20). These reflect the social upheavals in the 18th and 19th centuries as all Plains tribes adapted to incoming goods and technologies, resisted incoming settlers, their missionary agents and military protectors, and struggled with other tribes for diminishing territory and resources.

McCleary (2016) has recorded some recent interpretations by Crow people about this and other rock art sites in the region. This site is especially important because they see it as the core record of their history, and tribal members continue to make pilgrimages to the site (2016, 54). As in Lascaux, attention is frequently drawn to one particular group of figures, a complex palimpsest which appears to be a battle scene (2016, 123). However, Crow interpreters that Conner and Loendorf interviewed in 1972 treated it as a composite image drawn by three warriors over time, depicting their individual war deeds (McCleary 2016, 124). First one leader recorded his exploits, and then two others followed, each depicting their own deeds in response to the others. At the time of McCleary's own work, consultants extended this to four individuals (McCleary 2016, 128).

The differences in these various recollections do not mean that they are flawed from a social perspective, any more than the diverse interpretations of Lascaux. It is not the rightness that counts, but the fit among the community of interpreters. For the Crow, the interpretation of the petroglyphs is not the solving of a puzzle, but a timeless interaction between the images and the people who see them. As Roger Stops explained, "It's been so long now that we don't know who they are, but we can see what they had done" (Conner 1972, 137, quoted in McCleary 2016). And as McCleary emphasizes, these are histories because they are visible to be read. Elders are most often responsible for gaining the knowledge and interpreting the stories to their families, and the warriors counting coup, the dancers and the other images familiar to the Crow elders are stories to those who visit the rocks, and their meanings are in the telling of the stories, not in the details of their interpretations.

Rock art sites are thus not repositories of collective memory, as if each site holds specific contents, like a library or gallery. Narratives change naturally as the generations pass, but if individual people continue to visit sites and relate stories about them, the knowledge remains present and vital. The rocks themselves help sustain the continuity of meanings by their nature, as their forms, weight and persistence give them a strong presence. This fundamental salience intensifies when people know that their ancestors remain among them and their pictures are still there to affirm believed and recalled pasts. The rocks speak through these displays, urging the descendants of the painters and carvers to relate, recall and carry on. This convergence of physical and social energies strengthens and sustains the whole community.

If there is a reason to consider the grounds and situations of art and ideology in ecological terms, it is this direct engagement with the sources of attention. Overviews may provide insights, but just as the Paleoindian hunter who climbed the Tower to look for game in the distance had to climb down to hunt it, analysts must also come back down to the ground, where the world works, and discover them in action.

References

Bahn, Paul G. 1995. The Impact of Direct Dating on Palaeolithic Cave Art: Lascaux Revisited. *Anthropologie* 33:191–199.

Bandi, Hans-Georg, Henri Breuil, Lilo Berger-Kirchner, Henri Lhote, Erik Holm and Andreas Lommel. 1961. *The Art of the Stone Age: Forty Thousand Years of Rock Art*. New York: Crown Publishers.

Barham, Lawrence S. 2002. Systematic Pigment Use in the Middle Pleistocene of South-Central Africa. *Current Anthropology* 43:181–190.

Boas, Franz. 1911. *The Mind of Primitive Man*. New York: The Macmillan Company.

Brand, Adam (recorder). 1698. *A Journal of the Embassy from their Majesties John and Peter Alexievitz, Emperors of Muscovy, &C. Over Land into China, Through the Provinces of Vstiugha, Siberia, Dauri, and the Great Tartary, to Peking, the Capital City of the Chinese Empire. By Everard Isbrand, Their Ambassador in the Years 1693, 1694, and 1695. Translated from the Original in High-Dutch, Printed at Hamburgh.* London: Printed for D. Brown at the Black Swan and Bible without Temple-Bar and T. Goodwin at the Queens Head over-against St. Dunstan's Church, Fleet Street.

Bravery, Benjamin D., James A. Nicholls and Anne W. Goldizen. 2006. Patterns of Painting in Satin Bowerbirds Ptilonorhynchus violaceus and Males' Responses to Changes in their Paint. *Journal of Avian Biology* 37:77–83.

Breuil, Abbé [Henri]. 1950. Lascaux. *Bulletin de la Société préhistorique de France* 47:355–363.

Breuil, H. 1952. Lascaux. *Man* 52:110–111.

Brumm, A., A.A. Oktaviana, B. Burhan, B. Hakim, R. Lebe, J. Zhao, P.H. Sulistyarto, M. Ririmasse, S. Adhityatama, I. Sumantri and M. Aubert. 2021. Oldest cave art found in Sulawesi. *Science Advances* 7: eabd4648.

Campbell, Joseph. 1949. *Hero with a Thousand Faces.* New York: Pantheon Books.

Campbell, Joseph. 1960. *The Masks Of God: Primitive Mythology.* London: Secker & Warburg.

Chalmin, Emilie, Michel Menu and Colette Vignaud. 2003. Analysis of rock art painting and technology of Palaeolithic painters. *Measurement Science Technology* 14:1590–1597.

Conner, Stewart W. 1972. Interview with Chester Medicine Crow, Francis Bear Claw, and Roger Stops, June 10. Transcription in *Joliet Site File*, in the possession of Stuart W. Conner, Billings, MT.

Dart, R.A. 1974. The Waterworn Australopithecine Pebble of Many Faces from Makapansgat. *South African Journal of Science* 70:167–169.

Davenport, Demorest, and Michael A.Jochim. 1988. The Scene in the Shaft at Lascaux. *Antiquity* 62:558–562.

Dewdney, Selwyn, and Kenneth Kidd. 1962. *Indian Rock Paintings of the Great Lakes Region*. Toronto: University of Toronto Press.

Dewdney, Selwyn, and Kenneth Kidd. 1967. *Indian Rock Paintings of the Great Lakes Region*. Second Edition. Toronto: University of Toronto Press.

Dusenbery, David. 2002. *Sensory ecology: How Organisms Acquire and Respond to Information*. New York: W.H. Freeman.

Eliade, Mircea. 1964. *Shamanism: Archaic Techniques of Ecstasy*. London: Routledge and Kegan Paul.

Endler, John A. 2012. Bowerbirds, art and aesthetics. Are bowerbirds artists and do they have an aesthetic sense?*Communicative and Integrative Biology* 5:281–283. doi:10.4161/cib.19481.

Endler, J.A., D.A. Westcott, J.R. Madden and T. Robson. 2005. Animal Visual Systems and the Evolution of Color Patterns; Sensory Processing Illuminates Signal Evolution. *Evolution* 50:1795–1818.

Fodor, Jerry A. 1975. *The Language of Thought*. New York: Thomas Y. Crowell.

Gibson, James J. 1978. The Ecological Approach to the Visual Perception of Pictures. *Leonardo* 11:227–235.

Glory, A. 1961. *Le Brûloir de Lascaux Gallia préhistoire* 4:174–183.

Henry, Francois. 1941. Discovery of a Cave with Prehistoric Paintings at Lascaux (Dordogne). *Journal of the Royal Society of Antiquaries of Ireland* 11:61–63.

Henshilwood, Christopher S., Francesco d'Errico, Karen Loise Van Niekerk, Luca Pollarolo, Laure Dayet and Alain Queffelec. 2018. An Abstract Drawing from the 73,000-year-old Levels at Blombos Cave, South Africa. *Nature* 562: 115–118. https://doi.org/10.1038/s41586-018-0514-3.

Hoffmann, Dirk L., Diego E. Angelucci, Valentín Villaverde, Josefina Zapata and João Zilhão. 2018. Symbolic Use of Marine Shells and Mineral Pigments by Iberian Neandertals 115,000 Years Ago. *Science Advances* 4 (2): https://www.science.org/doi/10.1126/sciadv.aar5255.

Hoffmann, D.L., C.D. Standish, M. García-Diez, P.B. Pettitt, J.A. Milton, J. Zilhão, J. J. Alcolea-González, P. Cantalejo-Duarte, H. Collado, R. de Balbín et al. 2018. U-Th Dating of Carbonate Crusts Reveals Neandertal Origin of Iberian Cave Art. *Science*2018:912–915.

Jochelson, Waldemar. 1905. *The Koryak: Religion and Myths*. The Jesup North Pacific Expedition, Memoir of the American Museum of Natural History, New York, Volume VI part 1. New York: G.E. Stechert.

Kirchner, H. 1952. Ein archäologischer Beitrag zur Urgeschichte des Shamanismus. *Anthropos* 47:244–286.

Lewis-Williams, David. 2002. *The Mind in a Cave*. London: Thames and Hudson.

Loendorf, Lawrence, and Audrey Porsche. 1985. *The Rock Art Sites in Carbon County, Montana*. Grand Forks: Department of Anthropology University of North Dakota, Contribution #224.

Lower Brule Sioux Tribe. 2017. *Map of Lakota Place Names within the Great Sioux Nation Lands of the 1868 Fort Laramie Treaty*. Lower Brule: Lower Brule Sioux Tribe.

Maidagan, Diego Garate, and Oscar Moro-Abadía. 2012. Faut-il Renoncer au « style I » de Leroi-Gourhan? Une Révision Critique des Débuts de L'Art Paléolithique. *Bulletin de la Société Préhistorique Ariège-Pyrénées* LXVII:67–79.

McLuhan, Marshall. 1964. *Understanding Media: the Extensions of Man*. New York: McGraw-Hill.

McPherron, Shannon P., Zeresenay Alemseged, Curtis W. Marean, Jonathan G. Wynn, Denné Reed, Denis Geraads, René Bobe and Hamdallah A. Béarat. 2010. Evidence for Stone-tool-assisted Consumption of Animal Tissues before 3.39 million years ago at Dikika, Ethiopia. *Nature* 466:857–860.

Mithen, Steve. 1999. *The Prehistory of the Mind: The Cognitive Origins of Art, Religion and Science.* London: Thames and Hudson.

Molyneaux, Brian L., Nancy J. Hodgson and Rachel M. Hinton. 2000. *The Archaeological Survey and National Register Evaluation of Devils Tower National Monument,* Crook County, Wyoming 1997–1998. National Park Service Contract No. 1443-CX-1230–1297–00l. Vermillion: Archaeology Laboratory University of South Dakota.

Mourer-Chauvire, C. 1975. Les Oiseaux du Pleistocène Moyen et Supérieur de France. *Documents des Laboratoires de la Faculté des Sciences de Lyon* 64:1–2.

Ruxton, Graeme D., and Martin Stevens. 2015. The Evolutionary Ecology of Decorating Behaviour. *Biology Letters* 11: https://doi.org/10.1098/rsbl.2015.0325.

Schaedelin, Franziska C., and Michael Taborsky. 2009. Extended Phenotypes as Signals. *Biological Reviews* 84:293–313.

Simpson, J., and E. Weiner, eds. 2009. *Oxford English Dictionary.* Oxford: Clarendon Press.

Sundstrom, Linea. 2004. *Storied Stone: Indian Rock Art in the Black Hills Country.* Norman: University of Oklahoma Press.

Texier, Pierre-Jean, Guillaume Porraz, John Parkington, Jean-Philippe Rigaud, Cedric Poggenpoel, Christopher Miller, Chantal Tribolo, Caroline Cartwright, Aude Coudenneau, Richard Klein, Teresa Steele and Christine Verna. 2010. *A Howiesons Poort tradition of engraving ostrich eggshell containers dated to 60,000 years ago at Diepkloof Rock Shelter,* South Africa. Proceedings of the National Academy of Sciences/https://doi.org/10.1073/pnas.0913047107.

Van Schaik, Carel P. 2016. *The Primate Origins of Human Nature.* Hoboken: John Wiley & Sons.

5 How rocks speak

Ecology and rock art

The fascination with rock art rests on a stark contradiction. These painted or carved images are literally set in stone, often with little change over millennia; at the same time, however, they have no direct links with any particular individual or society. Life happens around them.

The artist's work conveyed the original meaning of an image, relevant to a family and community until, like all memories and practices, it began its slow change. The picture, however, continued on and persists now, long after the circumstances of its creation.

The study of the art has always been a push and pull between formalism and contextualism, from what pictures might convey in their forms to their wider relations in a social environment. From the late 20th century on, contextualism has become dominant (see Molyneaux 1977). In the subsequent history of rock art research, increasing interest in indigenous knowledge, combined with notions of the relativity of all knowledge, has led to a decline in archaeological thinking in favour of latter-day history and contemporary experience.

The problem is that the latter days never end, and increasingly complex interpretations of meaning and function seem to be the solution for the absence of empirical evidence. Thus, while Dewdney canoed along the painters' trails in the Canadian Shield (Dewdney and Kidd 1962; 1967) and, with his rice paper tracings, tried to follow their hands and get a sense of their surroundings through direct experience, more recent efforts follow the lead of philosophers, sociologists and anthropologists, and have a different artist in mind. This one is an agent or proxy who paints or carves in silent service to mind and society, conceived as some larger organic representation or, more recently, one that shifts shape endlessly. Such representationalism, which emerges in discourse, is difficult to understand next to a painted rock or among the people in an indigenous community today. Perceiving and recording images in fieldwork is difficult enough (see Colson 2011a; 2011b).

The choice to study the fossilized event, the analogue image, is a form of resistance to the idea of governing representations. It remains curious about the motivations in a social group for making images – especially the first in a

DOI: 10.4324/9781351053105-5

place – and the physical, emotional, aesthetic and ideological aspects of the act (see Arsenault's (2013) study of petroglyphs at Qajartalik, Nunavut). And in the search for evidence, it chooses to work at the varying scales in the social and material surroundings of the artist rather than down from representations.

If a rock art image has any past at all, it must be at the scale of its production, as it exists to a painter or carver who lived somewhere else, traveled to the rock, climbed, crawled, stood or sat at the face and worked. These acts describe an ecology, as each artist faces and resolves cascades of situations within the environment to carry out the work.

With only empirical evidence to go on, it seems appropriate to shift attention from the image as a central preoccupation to the situation of its production. By including the body, sensory systems and environment in the action, it might be possible to gain more insight into what the maker intended by the outcome we see on the rocks. Before Gibson developed his theory of ecological perception and Merleau-Ponty discovered the body, E.B. Holt made this comment about interiorization in general (Holt 1915, 55).

> … in order to understand what the organism is doing, you will just miss the essential point if you look inside the organism. For the organism, while a very interesting mechanism in itself, is one whose movements turn on objects outside of itself, much as the orbit of the earth turns upon the sun; and these external, and sometimes very distant, objects are as much constituents of the behavior process as is the organism which does the turning.

The case studies that follow, extending over two chapters, concern the question of how individual people might have acted in *their* circumstances. In rock art, they did not simply appear at a site; they traveled there, painted or carved their pictures one by one and then moved on.

In the discussion below, the rock art is important but only incidental, as it marks journeys with other motivations and destinations clearly outside the spell of its governing representations.

A painter's journey

The Canadian Shield is a region of exposed, weathered and glaciated Precambrian granite bedrock almost the size of Europe, with a subarctic climate, a daunting terrain of lakes, impassible muskeg bogs and dense forests, and an ecology mainly limited to hunting and gathering. The indigenous inhabitants today are a relatively small, widely distributed population speaking Algonkian languages.

Our interest here is in rock paintings recorded in the ecotone between the boreal forests and the mixed forests, parklands and prairies that extend from Lake Superior to the Lake of the Woods on the north, between the Mississippi River on the east, and the Red River and Big Sioux River on the west. Much of the environment has not changed materially since the time of the paintings, the latest being in the late 18th or early 19th century, and so the sites are much the same as they were when the painters worked there. Selwyn Dewdney conducted the seminal recording in the mid-20th century (Dewdney and Kidd 1962; 1967), and research continues to the present day. It is a daunting task for all those who study it in the field. Almost all the art is distant from roads on granite rocks that fringe the lakes, and because paintings were usually done from the water, they are only accessible by canoe or boat.

As the first consideration when thinking about the situation of the artist in the Shield is the journey (e.g., 2013), we begin this discussion by using a map. A Cartesian map will not do, as it is important to understand what it is like to navigate through the lakes and rivers by canoe and portage. This was, after all, the painter's experience. About 1730, the French trader Pierre Gaultier de Varennes, Sieur de La Vérendrye, sought to find a route to the western sea that was somewhere west of the Great Lakes. After journeying through the known territory he reached the west end of Lake Superior; he obtained from his Cree guide, Ochagach, a map of the water route from Lake Superior through the forests and waterways to the Great Plains.

This map will provide basic orientation among the sites discussed below. Most importantly, the way Ochagach conceives the environment provides an important clue as to why painters selected particular rock outcrops to paint among the myriad rocks they passed.

At first glance, the sketch looks like the internal organs of some fabulous beast, and seems impossible to use, as it lacks the proportions, relative distances and features of a familiar topographical map – but these do not matter here. The map comes directly out of Ochagach's experience following the route at the stern of a birchbark canoe. It depicts what the traveler needs to know from the water, just like navigating on the London underground or the New York subway or any bus or train route: the passage from one place to the next, one station, one lake.

The first European traders followed the Kaministikwa River route, which arches across the eastern side of this map. What Ochagach showed La Vérendrye was a more direct route to the west, following a chain of 25 lakes

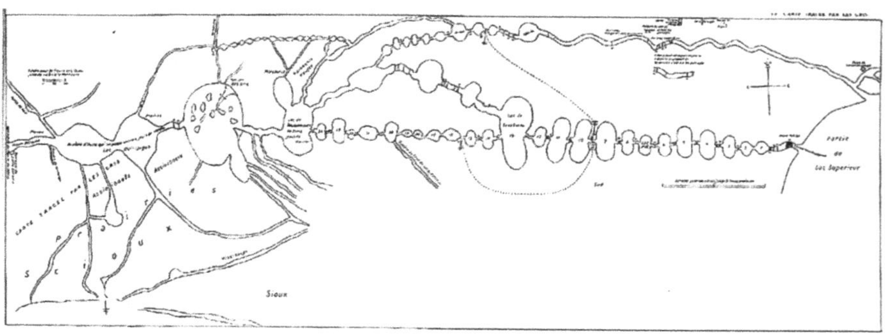

Figure 5.1 Ochagach's Map

between Lake Superior and the Lake of the Woods (visible here with its scatter of small islands on the western side of the map).

Following this route by reckoning from the water is a difficult experience, as over the great mass of the Shield the view almost never varies: angular, irregular lakes and waterways edged by the fractured faces of granite hills and forests. Ochagach's map depicts a sense of the task ahead, a steady paddle west, lake by lake, and it assumes that travelers will know what they will face along the way. The passage between lakes is often through a narrows or stream that is often hidden at the end of one of many bays, and because the elevation of the land rises and falls over the route, they may be impassable by water because of waterfalls, rapids or shallows, or because there is no water at all. In these circumstances, travelers must carry the canoes and all supplies by hand – portage them – overland between the waterways.

Portages are endured. Canoes and supplies must be carried up and down a trail that may be steep and strewn with rocks, thick with undergrowth, hot, humid and bug-ridden in the summer, and long. At the debouche of the western streams into Lake Superior on this southern route, the final stretch of river, at the east end of Ochagach's map, was what the Ojibwa call Gichi Onigamiing (great carrying place). This was a difficult portage more than eight miles long around the rough water, especially hauling large freight canoes and supplies, and, as the fur trader Alexander Henry would report 70 years later, there would be another 38 portages between Lake of the Woods and Lake Superior – a total of more than 18 miles (Coues 1897) – but it was more direct, and quickly became the primary route for the fur trade.

This is where the unrelenting panorama complicates the journey. Finding the portages without a map is often difficult. Lakes are irregular in size and shape, with numerous bays and islands. Without the advantage of a detailed view from the sky, the paddler who steered the canoe only a foot or so above the water had to use landmarks along the route.

In a society without maps and compasses, landmarks must be conspicuous enough to be described, passed on and recognized by other travelers following the route. This was a simple matter in the seas off Sri Lanka and at Grey Horn Butte. Sailors and tribes had their own accounts, but the descriptions converged on distinctive features recognized by all. From a canoe, low on the surface of the water, however, the only horizon is often a continuous flow of cliffs and smooth rock shorelines and forest edges, and so conspicuous rocks or other features along the way served a similar purpose. This would have been Ochagach's experience. He did not need the map he made for La Vérendrye, as he followed the path of his own senses and recollections along the route.

An example of the convergence of journey, place and description is a rock art site on Darkwater Lake, a few miles off the main route at the midway point on Ochagach's map between Lake of the Woods and Lake Superior. The outcrop is at the head of a small bay leading to a portage into a much larger lake. The rock is a huge, sheer cliff, heavily streaked with blue-green algae and traversed by large fractures running at a steep angle down the cliff face. At the base, the largest plunges almost to the water's edge and then ricochets sideways, forming a nearly horizontal crevice.

The paintings are in bare patches of rock between the huge sheets of algae and lichen. Just above the horizontal crevice, a painter arranged a group of six canoes and a moose above a horned serpent, whose tail and head curve up like a bracket around them.

In Ojibwa oral tradition, the serpent is *Mishipizhiw* (the Great Lynx), a spirit being that lives in the waters of the lakes and, without appeasement with an offering, may stir up these waters (e.g., Dewdney and Kidd 1962, 114), while the prominent crevice recalls the belief in the *maymaygwashiwuk*, the little people who live in rocks and, sensibly, use visible openings, such as cracks or crevices, to come and go. Some people believe they are the painters of the rock art.

For archaeologists and rock art researchers curious about how this site came to be, there is little more that can be said – but there is a practical message in this scene for all travelers, ancient and modern. It is sometimes important to pause and take care, because travel on the lakes can be dangerous. In the lakes and

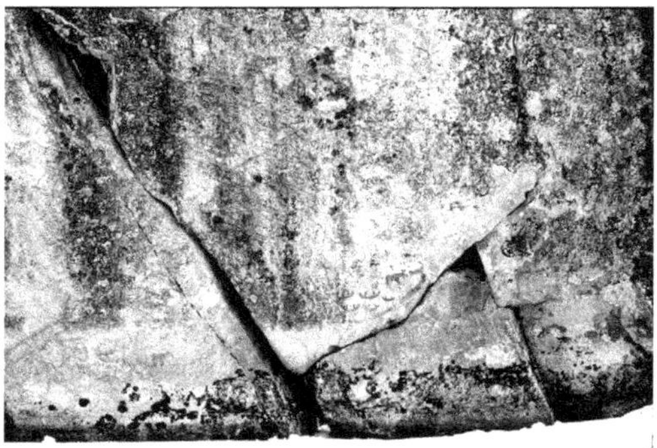

Figure 5.2 Darkwater Lake Rock Painting Site

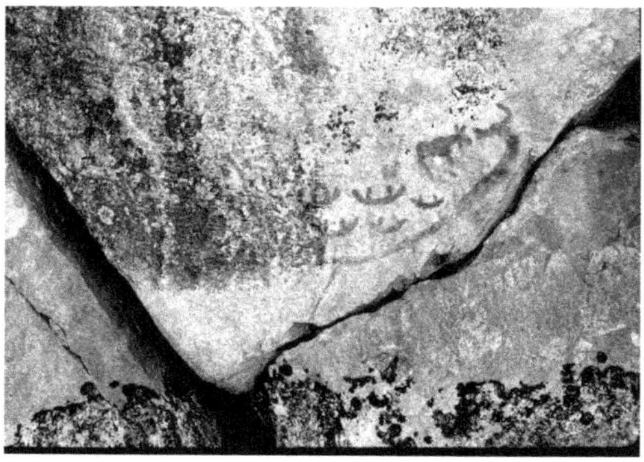

Figure 5.3 *Mishipizhiw* and Canoes

rivers of the Shield, the pleasant glide of a canoe over calm water can end quickly, and when the winds are high they can force the bow into the troughs of advancing waves, or create a treacherous chop near the shore, as incoming waves crash into the waves rebounding from the rock cliffs.

So, if it is impossible to know what the artist meant in painting this scene, it seems clear that as the painter and the archaeologist both canoed to the site hundreds of years apart, the act of painting suited the journey and the subject fit the experience, even if the knowledge of the creature as *Mishipizhiw* is confined to the Ojibwa.

This convergence of effort suggests a different way to look at the interpretive problems of rock art – applying a working vision of the past rather than a spiritual one, in which the focus of interest is the painters as flesh and blood rather than mere vehicles for symbolism.

This is difficult to maintain, because the lure of the imagination is so compelling. Working from artifacts and rock art records alone, object and form dominate environment and context, and object-bound approaches continue to encourage the overwhelmingly ideological interpretations of the past today. To be sure, the encounter at a site is no different. The salience of the images strikes the senses, emotions and intellect in a single blow.

The problem with following the lead of an image into what it might represent is that it does not respect the challenge of understanding the art as it emerged in its surroundings, in the work of a painter who traveled through the rock-lined lakes to the site, and the practical questions about how this painter happened to come upon this rock in the first place to paint and why.

To take this spare, but more integrated view, it is necessary to ignore the classifications that dissolve experience into products of the imagination – the general assumptions, models and representations meant to account for individual behaviour – in favour of what happens at the local level, where people, minds and environment interact.

To make the point, we will cast out the idea we have begun to develop, the unity of people, canoes and rock art among the waterways of the Shield – even though it is the most popular impression of rock art today. The reason is that it is easy to get lost in a terrain of ideas and forget about the ordinary circumstances of life – such as the weather. Weather, as opposed to climate, rarely enters into consideration in archaeology because in the realm of ideas and processes, there is no weather to speak of.

Between November and April, nightly low temperatures in the southern Shield gradually drop below freezing, ice forms on the lakes and canoes go into storage. This is a different world. In the long winters, outsiders have gone south, inhabitants spend most days around their homesites, except when hunting or fishing, and travel can be arduous and dangerous on foot, by dogsled or, today, by snow machine.

This suggests that the world of rock art is barely a half-life – but what if the habits of the painters simply changed with the seasons? Winter can create favourable conditions at rock faces along the shores of frozen lakes. Ice may heave up along shorelines and, along with snow drifts, provide a higher and much more stable platform than the bottom of a canoe. And the experience of working at a site in late winter can be surprisingly pleasant. When sunlight begins to intensify in March and the winds are calm, it can be warm enough that water from melting snow and ice drips from above the rock face. The Darkwater Lake site was recorded at this time in these conditions in 1985. For certain, we know that the whole sense of travel changes. Now the impassable muskeg is frozen, portages are gone and routes overland through the forests may be shorter than the meanders of the rivers and lakes.

It is appropriate that perhaps the first winter rock art site recorded in the Shield was through the efforts of Phil Sawdo, an Ojibwa trapper who lived in the Quetico region of northwestern Ontario (for an account of his life, see Sawdo 2004). He knew of a small group of rock paintings well above the margins of the nearest lake (north-northeast of Lac de Sesakinaga (Saganaga Lake) on Ochagach's map), and the location distant from water seemed unusual. Getting there by water was a 40-mile trek by canoe and portage through a chain of small lakes around a large land mass known as Hunter Island, and then a hike upland through thick black spruce to the site (Molyneaux 1986). I speculated at the time that it might have been done during a hunt in the interior from a camp along one of the lakes – until Mr. Sawdo explained that it was along his old winter hunting trail, a route that crossed frozen creeks and muskeg in an almost straight line from the canoe landing. This route, only passable on frozen ground, cut off more than half the journey and, on snow machines, took hours instead of days, as we found following the overland route the next winter. The whole Shield has trails that once joined its scattered inhabitants in winter life, and these paintings, and perhaps others, are its records. Recently, in a study of rock art in the Temagami area of northeastern Ontario, Zawadzka (2013) noted that the local Teme-Augama Anishnabai population had several *Bon-ka-nah* (winter trails) near rock art sites (2013, 183).

Spending more time in the field or at the source of any data and thinking about the situation of production brings more life into the analysis. Back at Darkwater Lake, if only the images count, it does not matter whether or not the artist intended to depict an actual or metaphorical journey. The direction of attention in the analysis has no mediation in experience. Searching always upwards leaves the evidence of individual acts and their outcomes behind, and causes the artists, their work, their territorial bonds and their communities to dissolve into the background – as if they had no part in the creation of the art.

To consider the grounds of rock art, it is necessary to rethink the idea that human acts are the work of collective forces, such as shared culture, which reside in ideology. There are alternatives to social group theories that are inimical to institutional thinking but can conjure up similar patterns. If we turn to the world as a model instead, the painters in the Shield and their journeys exist in a discrete system of organization, sharing a kind of emergent collective behaviour.

Such decentralized behaviour is evident in such collectivities in animals as clouds of insects, murmurating starlings or herds of bison (e.g., Giardina

2008). It is the specific interactions of each individual with others in the common environment that yields the appearance of the group. Among humans, collective behaviours may be animated by shared ideas or by more discrete, physical forces that act collectively. Simple physical urges and imperatives such as those involved in navigation, or getting water and food, create patterns of necessity that do not require the cultural rationales that participants and analysts might offer. Thus, over millennia, hunter-gatherers, horticulturalists and urban dwellers follow similar paths through terrain, heed similar landmarks along the way and converge in pleasant ground near the shores of lakes and rivers without prior instruction.

A way to explore the field of action possibilities around a rock art site without the incursion of representations is to imagine the site without any images at all – the rock that the first artist saw. The choice of a particular bare rock to paint would then have to relate to factors outside the site – aspects of the locality that could be either visibly or ideologically salient (Molyneaux 1983). One cannot simply assume, therefore, that several centuries ago the act of the first painter was to anoint a sacred site. The picturing of a spirit being, or an invocation, or a prayer spoken before or during a journey across the waters where it lived, does not automatically render this part of the world special and different from the rest. The acts of the painter may have consumed an hour in a whole life and were probably not intended as timeless evocations of a fundamentally spiritual world.

Thinking of even the rigour of a single day, rock art studies need to do more than consider images as if they were painted or carved as symbols within landscapes of a culture's own devising. Artists in the Shield, like anywhere else, shared technologies and metaphorical resources within their social groups – hence, the existence of common subjects and styles over space and time – but this does not mean that they were governed by collective memories, values and experiences. The natural topography has always afforded the possibility of overviews. The hunter who camps beside the Tower climbs its slope to see up and down the river valley, and then returns; the pilgrim walks the many steps to the Peak and gazes over the lands and sea, and the traveler in a lifetime of canoeing through a homeland of lakes and streams, climbs to the top of a high granite cliff on Beaverhouse Lake and daubs some red ochre pigment there (Molyneaux 1987), but they all must go back down to continue their lives.

At Darkwater Lake, there is a reason to ponder this problem of representation and the significance of the journey. To the right of the serpent and canoes and painted by a similar hand is a figure of a hunter who fires a gun. It is a musket; the painter has added a daub ahead of the barrel to represent the burst of black powder smoke.

Two longer treks

In the early 18th century, during tensions between the Ojibwa and Sioux in the southern Shield and adjoining parklands and prairies in present-day

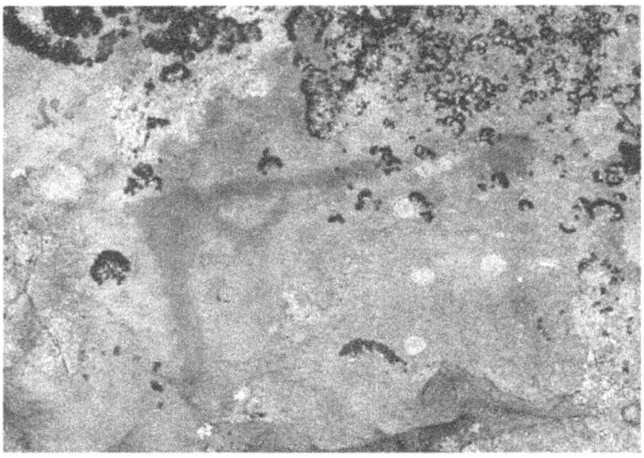

Figure 5.4 Hunter Firing Musket, Darkwater Lake

Minnesota, it is said that a party of Sioux warriors traveled from their prairie homelands into Ojibwa country, and at a rock painting site on Crooked Lake in the Shield they fired or pushed arrows into its crevices as a hostile act.

At an unknown time, but probably during the same period, it seems that some Ojibwa traveled from their forest homelands into Sioux territory in southwestern Minnesota and left a scatter of petroglyphs there.

Minnesota has a complicated geography because of its position near the southern extent of glaciation in North America over the past two million years. The advances and retreats of sheets and lobes over thousands of years from the north, east and west have scraped the surface or deposited drifts of till, moraines and other distinctive features in their wake. The most recent advance was 14,000 years ago, when a narrow glacial lobe flowed diagonally northwest to southeast across the southern third of the state, creating the environment that exists in the region today (Leverett and Sarde 1932; Bray 1977).

Once the Laurentide ice sheet began to melt back northwards, just below the southwest corner of the state, where the new course of the Missouri River turned southwards, indigenous people were already recolonizing the succession of tundra, forests, parklands and prairies behind the retreating ice, as the surface find of a Clovis point (13,000–10,000 BCE) on a ridge of wind-driven silt near the mouth of the Big Sioux River in Iowa suggests (Molyneaux 1998). By 9000 BCE, the ice and the lake had cleared all the lands south of the Lake of the Woods (Dyke 2004).

The pattern of drainages in postglaciated lands set the patterns for the movement of people, as they foraged and traveled by foot. The northern part of the state, which is at the southern edge of the Canadian Shield, has a divide between lakes and streams in the northeast that flow into the Great Lakes and eventually into the Atlantic and those in the north-central and west that flow through the Rainy River watershed into Lake Winnipeg. Most of the rest of the state's waters flow into the Mississippi River, which has its headwaters south of the Rainy River drainage, except for the western margins along the edge of the Great Plains, which either flow north into the Red River valley or, in the southwest corner, into the Big Sioux River, which debouches into the Missouri River and, eventually, into the Gulf of Mexico.

This diverse region offered more affordances for hunting and gathering people than the Shield proper: a temperate climate, rich soils, wide range of plants and animals and a topography of dissected streams and lakes that allowed efficient travel in all directions. It has a rich and complicated archaeology, reflecting influences from all directions over time.

Unfortunately, the ease of movements also made the whole region vulnerable to invasion. This was especially so from the east, as the postglacial watersheds were not equal. The Great Lakes was a huge, open front for rapid movement unmatched by the meandering rivers in other directions. This allowed European colonizers to move quickly from the eastern coastal fringes into the interior in the 17th century, carrying goods, technologies, ideas and diseases that undermined thousands of years of mutual adaptation and interaction among indigenous peoples.

The most dynamic force in this colonization was the fur trade, which quickly became a catalyst for the expansion of the largest indigenous societies in northeastern North America, the Iroquois and Ojibwa. Once these tribes obtained guns and, in the competition to exploit the new sources of wealth and technology, gained control of large territories, they were strong enough militarily and economically to wield power in relation to French, British and Dutch colonists and traders. The expansion of both groups, and the ability of the Ojibwa to resist the Iroquois (Eid 1979), compelled smaller societies to move west, where they came into increasing contact with Siouan speaking peoples who lived in the forests, parklands and prairies south of the Shield.

The result in historic times was a bewildering mix of highly mobile indigenous societies (Hämäläinen 2019). By the end of the 17th century, the Ojibwa occupied the area around the western end of Lake Superior, and the Sioux lived in the mixed forests and prairies south of Lake Superior and west to the Great Plains. By the end of the 18th century, however, the Sioux had obtained horses and quickly expanded into the Great Plains to take advantage of the vast herds of bison, the Ojibwa continued their migrations west and south into the southern Shield and other tribes moved west as the Iroquois expanded in the northeast. In this complex material and social terrain, indigenous and outside societies negotiated their lifeways and interests among forests, grasslands and waterways, always in relation to colonial power and the lure of fur trade – and, in a few places, some of these dynamics are reflected in rock art.

Crooked Lake

In 1789, as the Scottish explorer Alexander Mackenzie followed the southern water route between Lake Winnipeg and Lake Superior that Ochagach had sketched out for La Vérendrye about 60 years earlier, he noted a curious rock.

> Within three miles of the last Portage is a remarkable rock, with a smooth face, but split and cracked in different parts, which hang over the water. Into one of its horizontal chasms a great number of arrows have been shot, which is said to have been done by a war party of the Nadowasis or Sieux, who had done much mischief in this country, and left these weapons as a warning to the Chebois or natives, that, notwithstanding its lakes, rivers, and rocks, it was not inaccessible to their enemies.
>
> (Mackenzie 1801, liv)

When David Thompson passed by on August 17, 1797, while surveying the disputed boundary between Canada and the United States for the North West Company, he provided more details of the story (Tyrrell 1916, 178):

> A short distance south eastward of the Height of Land in the crevices of a steep rock, about twenty feet above the water of a small Lake, are a number of Arrows which the Sieux Indians shot from their Bows; the Arrows are small and short. The Chippaways, the Natives say these Arrows are the voice of the Sieux and tell us, "We have come to war on you, and not finding you, we leave these in the rocks in your country, with which we hoped to have pierced your bodies." This was about the year 1730.

In his pioneering surveys of rock art sites in the Shield, Selwyn Dewdney recalled his first sight from a canoe: "a great bulk of granite leans ominously over the water, its walls streaked with a rich mosaic of iron stains, vari-coloured lichens, and vivid deposits of precipitated lime" (Dewdney and Kidd 1962, 30).

Some of the arrows were still there in 1823. Major Joseph Delafield, a member of the commission negotiating the boundary between the United States and Canada along this waterway, noted that "the fissure presents an opening of two inches, and there may be seen still the feather ends of about twenty arrows driven nearly to the end" (Nute 1948, 133). Some of the projectile points were eventually recovered from the site in an archaeological survey (Birk 1974).

Without the story of the Sioux, this striking landmark would be mapped, recorded and interpreted as an Algonkian rock art site, and the arrows associated with a ritual act. Indeed, the interpretation could draw from similar evidence at a rock painting site on Lower Manitou Lake, about 90 miles northwest in the Lake of the Woods region, where a small, triangular projectile point lay on the bottom of the lake below the paintings (Molyneaux 1983), and also observe that Ojibwa people continue to offer to the spirits in rocks which have paintings. With the stories that Mackenzie and Thompson picked up, however, it is not possible to dispense with the site so easily. The firing or placing of the arrows in the rock may have been a political act.

During the 1730s, the Dakota Sioux inhabitants of the region pushed back against incoming Ojibwa in the lands south and west of Lake Superior, and there were notable skirmishes across the edge of the Shield. The Sioux choice of this site on Crooked Lake for their political statement would have been strategic, as this was a striking landmark along the primary fur trade route. And if they knew the rock paintings were nearby, the impact of their arrows may have been much greater. Because of their salience and the messages they may convey, images have been frequent targets of political action, and rock art in North America cannot be an exception. On Crooked Lake, however, there is no evidence of defacement and, in fact, almost no evidence that anyone traveling through the region at the turn of the 19th century even saw the rock art. Among the early visitors, only Major Delafield in 1823 noted any markings on the rocks (Nute 1948, 132), and he wrote, cryptically, that "the traveler has left his mark on this rock in various ways; some by name & some by date & some by strange device" (Delafield 1943, 415). As Nute observed, the "strange device" probably referred to the rock paintings nearby. Thus, Delafield saw images, but no art, and his fellow travelers saw nothing at all. Dewdney's description of the rock art may be the best explanation: "Here man's art is apt to be unnoticed, modestly appearing some fifty yards south of this colour display" (Dewdney and Kidd 1962, 30). From this, it is possible that the natural salience of the great rock overwhelmed the Crooked Lake rock paintings and the Sioux warriors may not have noticed them at all.

There is another issue related to this story, which relates more fundamentally to the relative significance of the rock art of the time. If the Sioux did fire the arrows and if, by all accounts, great landmarks and rock art sites are sacred places, why did the Ojibwa not remove them in the many decades between the Sioux incursions and Delafield's visit in 1823? Here we confront the abyss, where only ideologies reside. Today, rock art is a scarce resource, and it is most significant to the descendants of its makers, who draw history and belief from it. When image-making was common practice, however, the painters who set their canoes towards some great rock had memories and relations that emerged from *their own pasts.* Their motives are no doubt within the legacy that their descendants hold today, but they could not anticipate a future in which they no longer lived freely among these waterways or followed their own traditional ways.

If there is any more of an answer to the Rock in Arrows on Crooked Lake, it will not be found by pondering the images. It is there in the traces of life along the cliffs and shorelines, with every patch of pigment and flake of stone or fragment of pottery that survives from this time. Whether or not travelers, supplicants or combatants left the arrows or painted nearby, research should be as much bound to the stony terrain as to general ideas and, as Ochagach did with his map, chart the journey.

Pipestone and Red Rock Ridge

The idea of the Ojibwa traveling outside the closed waterways of the Shield and their familiar grounds is unremarkable, as in both oral history and the evidence of the recent centuries the tribe has always ranged widely across the eastern forests. It is less usual, however, to see visible evidence of movements southwards into the prairies and plains beyond the Shield. One clue in the southern Shield is a site on the east side of the Lake of the Woods, along the rocks of Whitefish Bay, where a group of red ochre paintings of bison surrounded by handprints faces the water from a sheer granite rock. This is a rare subject in boreal forest rock art. The nearest bison lived in the open forests and grasslands of present-day Minnesota before their extirpation in the 19th century (Swanson 1940). The easiest water route from Whitefish Bay to bison pastures was a two- or three-day journey to the southwest end of the lake. If the trek was after about 1750 and Ojibwa hunters gained horses there, they could even continue overland towards the Red River valley, where great herds of bison roamed into the mid-19th century (Hargrave 1871).

Our interest here is some evidence of even longer treks, recorded in petroglyphs, pounded into a reddish quartzite bedrock in a terrain of glacial drift plains and wind-driven silt soils, 350 miles south-southwest of the Lake of the Woods region. The rock is mostly buried, but in several places in southwestern Minnesota and adjoining South Dakota it rises above the grasslands in conspicuous ridges, mounds and erratics, polished by exposure.

The defining feature in this region, and central to this discussion, is a rugged upland formed during the final Wisconsinan glaciation. About 14,000 years ago, two lobes formed, one moved south, covering the east half of South Dakota, and the other flowed southeast across the lower third of Minnesota. The gap between them formed a complex upland of moraines and till, a product of previous glaciations, which the French called the *Couteau des Prairies*. This couteau has two drainages – the Cottonwood River, which flows into the Minnesota River and then into the Mississippi, and the Big Sioux River, which flows south just outside the southwestern corner of Minnesota into the Missouri River. Both eventually debouch into the Gulf of Mexico.

In one place, at the edge of the Couteau about 360 miles south-southwest of the Whitefish Bay rock painting site, an exposed cliff of quartzite rises up 10 to 15 feet above the grasslands and runs for two miles before dipping back into the earth. The formation is slightly tilted, so that its layers dip into the ground, and at the bottom of the exposure is a layer of argillite, a clay siltstone, with a rich brownish red seam, coloured by hematite. This material is soft enough to work with hand tools.

The earliest evidence of quarrying at the site comes from analysis of artifacts found elsewhere. Many societies worked the stone into ornaments or tools, especially smoking pipes, or bartered for these valued objects, as they appear widely across the central part of the continent among horticultural societies, beginning early in the Woodland period – for example, stone pipes from Wisconsin, dated ca. 2500 BP (Boszhardt and Gundersen 2003) – and they are frequent in the mounds and settlements of later Hopewellian and Mississippian cultures in eastern North America (e.g., Brown 2006; Emerson et al. 2013).

When French fur traders arrived in the region, they called the pipes *calumets*, and it soon became clear among the Sioux and other indigenous tribes that the pipes, and their use, were highly spiritual, figuring in origin stories and central to religious ceremonies. With the evidence of the red stone artifacts in elite contexts in prehistoric sites and the importance of the smoking pipe, the notion of a widespread calumet ceremonial tradition developed, all focused on these colourful rocks in southwestern Minnesota (Blakeslee 1981). Over the course of European and American history in the region, Sioux tribes (the Sisseton and then the Yankton) have managed the quarries, while allowing the extraction of the stone by many others. The property became Pipestone National Monument in 1937.

George Catlin (1796–1872), an American adventurer and painter who was one of the first outsiders to visit the site, described the quarry in active use in 1836 and mentioned petroglyphs (1859, 646):

> ... over an extended plain are seen, like gopher hills, their excavations, ancient and recent, and on the surface of the rocks, various marks and their sculptured hieroglyphics – their wakons, totems and medicines – subjects numerous and interesting for the antiquary or the merely curious.

In a survey and inventory for the National Park Service, Scott et al. (2006) identified 79 distinct glyphs in the Pipestone area. Most of these are on small slabs that once formed an exposed and weathered floor at three large glacial erratics, which sit oddly on the prairie by the quarries. A local stone sculptor, Leon H. Moore, pried 36 of these slabs out of the ground in 1888 or 1889 (2006, 98). Most have characteristics of images found in other parts of the Great Plains, but a few strongly resemble Algonkian rock art (e.g., 2006, 262–264). They have features common in the Shield rock paintings and figures incised on Ojibwa Midewewin birchbark scrolls (e.g., Dewdney 1975), and they continue as a primary motif in modern Ojibwa art.

Fifty miles due east is another place where the reddish quartzite rises above the grasslands. In the Cretaceous, this was probably an island in a dank swamp, but now it is Red Rock Ridge, a low prominence above the plains with scattered rock exposures, running almost east–west over a distance of about 23 miles. There is no pipestone here but, in one place, where the surface forms an almost level floor, polished by thousands of years of weathering, thousands of petroglyphs are visible. The glyphs here reveal an art of great age and diversity: bison, hoofprints, salamanders, turtles, elk, human stick figures, birds, leather bags and various weapons (atlatls, spear points, arrowheads and lances). The hardness of the quartzite is such that glyphs could survive for thousands of years without visible weathering, and the presence of atlatls and absence of horses suggests that the work was done over an unknown period of time before the mid-18th century (Lothson 1976; Buhta, Anfinson et al. 2017). Many of the images resemble forms and subjects found further west in the Great Plains, but a few, just as at Pipestone, resemble Ojibwa or other Algonkian forms (Callahan 2001, 77).

So, what can we make of the glyphs in general and the possible presence of Ojibwa people in particular so far away from their homelands? The simplest answer is that they traveled south from the Shield to quarry or barter for pipestone, and stopped along the way at Jeffers – but there is still a curious imbalance between the two sites, one which suggests that the answer is not that simple.

On one hand, there is Pipestone. It is physically prominent, a long cliff with a running stream and falls, huge glacial erratics that stand nearby like sentinels, a strategic location on the Couteau near the Big Sioux River, close to its debouch into the Missouri River, and a buried layer of soft red stone, a material desired over thousands of years for ground stone pipes and implements. For all this, however, it has only a scatter of rock art.

On the other, there is the site on Red Rock Ridge. It is a quartzite bed exposed on a low ridge in grassland, surrounded by local prairie resources and small lakes and streams, with no cliffs, no soft red stone, no creek running by or waterfall, and yet it attracted people over millennia to pound thousands of images into its polished surface.

Given the attractive affordances of Pipestone, both material and social, the high regional mobility of peoples across the region over thousands of years that the petroglyphs at Jeffers reflect and, in general, the assumption that ancient art is sacred, it is reasonable to expect more petroglyphs at Pipestone – indeed, the balance could be worse, as the prairie sod may cover more glyphs at Jeffers and other places along the ridge (Buhta, Anfinson et al. 2017). What then gave the Red Rock Ridge site its lopsided importance? While neither art nor archaeology provides anything specific to go on, and the ethnohistoric record is completely dominated by Pipestone, the answer might lie outside these ideological considerations, in precisely how people living in the Shield, the Great Lakes and the Mississippi River basin found their way south and west to the quarries and beyond.

The journey to Pipestone

A single record of a journey by an indigenous traveler from the north in the early 19th century exists for Pipestone – the route that O-kup-pee, a Métis interpreter and guide, followed in 1836 as he led George Catlin and Robert Serril Wood by canoe and on horseback to Pipestone. The route was a well-established fur trade route by then, but the details of this trek create a field of possibilities for much more ancient movements of people in the region, before the horse replaced the dugout and canoe for travel.

The narrative of the trip is Catlin's, and it is scattered in short quips in his rambling accounts across two books (Catlin 1859; 1899). It begins at Green Bay on Lake Michigan, as Catlin and his English companion Robert Wood rode in large birchbark canoe paddled by "five French voyageurs" up the Fox River to Fort Winnebago, at the east end of a short portage into the Wisconsin River. There, they obtained a canoe and paddled up the Wisconsin to the Mississippi, destined for Prairie du Chien, a few miles upstream from the confluence (1859, 641–42). Then, in his words:

> After we entered the Mississippi we had some hard work against its boiling current before we reached Prairie du Chien; and from that to the Falls of St. Anthony, a distance of four hundred and fifty miles, some "up-hill work" again".
>
> (1899, 168)

They eventually arrived at Fort Snelling, located at the mouth of the St. Peter's River (now the Minnesota) just below the Falls of St. Anthony (now in downtown Minneapolis), on Wednesday, August 16, and left four days later (Taliaferro 1836).

At Fort Snelling, Catlin first mentions his guide, O-kup-pee (1859, 698) – we are not certain if he joined the two there, or several days earlier at Prairie du Chien. Now a company of three, they continued by canoe upstream on the St. Peter's River for 100 miles, to Le Blanc's trading post at Traverse de Sioux (1859, 647). M. Le Blanc (signifying "white man") was Louis Provençalle, a French Canadian fur trader; he occupied this post for 30 years with his Dakota wife, raised a large family, and as a Protestant missionary observed, "spoke the Dakota language with fluency and force, but with bold violations of grammatical rules" (Babcock 1939, 260–61).

Here, Catlin exchanged for horses (he mentions returning the horses at the end of the trip; 1899, 173), and the three crossed the river and followed a course upstream along its left bank, below Swan Lake (mentioned on their return trip; 1859, 709), and recrossed to the right bank at the Cottonwood River, about 30 miles above the mouth of the Blue Earth River, from whence they"steered in a direction a little North of West for the Coteau des Prairies" (1859, 699).

About this final trek, Catlin writes:

> For many miles we had the Coteau in view in the distance before us, which looked like a blue cloud settling down in the horizon; and we were scarcely sensible of the fact, when we had arrived at its base, from the graceful and almost imperceptible swells with which it commences its elevation above the country around it.
>
> (1859, 699)

Eventually, they reached a most important destination:

> our Indian guide brought us to the trading-house of an old acquaintance of mine, Monsieur La Fromboise (sic), who lives very comfortably, and in the employment of the American Fur Company, near the base of the Couteau, and forty or fifty miles from the Pipe Stone Quarry.
>
> (1859, 662)

LaFramboise called his post Grande Lisiere; the geographer Joseph N. Nicollet, who surveyed the region three years after Catlin, would write in his diary, that it was "the last isle of woods we are to encounter for three or four days" (Timmerman 2009, 25).

LaFramboise guided Catlin and his companions the rest of the way to Pipestone (1859, 663), which they approached from the east (1859, 654). He knew the quarry well, as he dug pipestone there to supplement his trade (Timmerman 2009, 26), and it was he who guided Nicollet in his detailed topographic survey of the Upper Mississippi region in 1839.

When Catlin first wrote about his adventures at Pipestone in 1841, readers across the world saw an engraving of his beautiful painting of the Blackfoot leader Stu-mick-o-sucks (the buffalo bull's back fat), which he had painted in 1832; and in his book of letters he had written during his adventures, they read his description of "a beautiful pipe, the stem of which is four or five feet long, and two inches wide, curiously wound with braids of quills of various colours, and the bowl of the pipe ingeniously carved by himself from a piece of red steatite of an interesting character"; and they followed his absorbing account of his journey to this site, which he had earlier described as "somewhere between this place and the Falls of Anthony, on the head waters of the Mississippi". He was certain that this marvelous stone had "but one origin in this country, and perhaps in the world" (1859, 61).

The private spirituality of the ceremonial smoking pipe that Catlin made visible to outsiders transformed the quarries for an outside audience into a locus for the popular view of these western lands – a world of adventure, among exotic people and customs, and with a tragic destiny. It was no surprise that a contact of Catlin's, the geologist C.T. Jackson, named the stone *catlinite* after completing its mineralogical description (Anonymous 1839).

When the harsher reality of the treatment of indigenous people in North America began to replace stereotypes in the later 20th century, the special nature of Pipestone and its catlinite continued through the efforts of archaeologists. From about 1700 BCE to the coming of Europeans several large and sophisticated horticultural societies rose and subsided in the larger river valleys of eastern North America. These groups were distinctive for their construction of huge mounds and other earthworks, and their desire for exotic raw materials, including meteoric iron, copper and silver, mica, obsidian, marine shells and shark and grizzly bear teeth, which they used in art and ornamentation and, pertinent here, a wide variety of finely ground and polished stone smoking tubes and pipes.

What might have been the circumstantial exchange of cultural materials, technologies and ideas across a region became, in some influential research in the later 20th century, a vision of a centralized system of trade in prestige goods, controlled by elites, which extended as far as the core cultural features and artifacts spread (see Emerson and Hughes 2001).

Catlinite was implicated in these constructs because red ground stone pipes and other objects scattered throughout sites of the cultures were commonly assumed to have come from its quarries. This view was greatly influenced by the presence of a very large area of scattered settlements on the Big Sioux River near Pipestone, known today as Blood Run. Visible occupations began there about 900 CE and it was intensively settled between 1500 and 1700 (Henning and Schermer 2004). The later prehistoric settlements had a Mississippian character, with evidence of hundreds of mounds and other earthworks and considerable pipestone from the quarries and, at first contact with the French in the 18th century, the inhabitants were members of the Omaha tribe (Henning and Theissen 2004).

When James Gundersen published the mineralogical description of catlinite (1991), however, archaeologists were able to reassess the sources of existing artifacts and, where they found them to be different, search for other sources. At the huge urban settlement of Cahokia, which prospered between the 11th and 13th centuries on the banks of the Mississippi across from the present city of St. Louis, Missouri, Emerson and Hughes determined that "the vast majority of materials important to prehistoric Cahokians could be obtained from the nearby Ozark Highlands or the adjacent southern Illinois chert quarries" (2001, 156).

Such discoveries did not diminish Pipestone as a source, but rather brought it down to the humbler evidence on the ground, along with the grand economic models and their representations of past culture. While most regional archaeologists accept the model of a hypothetical cultural group of Mississippian horticulturalists with its own sphere of influence, defined in the 1950s as Oneota (Willey and Phillips 1957), the direct evidence of excavations suggests that the ancestors of societies known generally today as the Ioway, Omaha, Ponca, Arikara and Oto used Blood Run "as a redistribution center for trade items, including red pipestone, throughout the Midwest" (Fishel et al. 2010, 168, citing Alex (2000, 204) and Henning (2007, 76–79)). In this regard, the distribution of catlinite probably resembles that of other highly valued materials, such as obsidian, which may circulate through pickup, curation and individual or small group transport from its source, without requiring economic rationales and coherent systems of trade and exchange (Molyneaux 2002).

For the indigenous inhabitants of the region, the idea of Pipestone has never been a problem, as their association with the place and the stone has been continuous over thousands of years, even if the unique relations of each society are more difficult to distinguish after almost 200 years of interference. From the beginning of the first encounters with European, American and Canadian traders and adventurers, indigenous people were lured into narratives they did not choose and, over the centuries since, cajoled and forced to act and make statements in public about private spirituality. With this in mind, the ceremonial traditions associated with the pipe, the protection of the site and the continued operation of the quarries help sustain indigenous relations and values and resist the torrent of ideologies from outside, despite the political and social difficulties this entails.

The route not taken

For the glyphs on Red Rock Ridge, there is nothing to say during the same period. Outside the regard that any indigenous inhabitants in this area had, and the visits of local settlers, there are no accounts, no images on the rocks of horses or other historic items, not a single reference before 1885 (Lothson 1976). Off the beaten track, the site lay undisturbed on its ridge, once in a Cretaceous sea and now isolated in a sea of grass. Lacking the significance and notoriety of Pipestone, it would not have a future as a national monument, despite the international importance of its rock art and, to this day, remains a regional attraction.

What if Catlin had seen the glyphs here? He might have been attracted by the great diversity of curious images and sketched a few, and it might have even influenced his work at Pipestone, but probably not. His actions at the quarries showed that, of all these sights he saw, the one feature that did not capture his interest was the rock art. He mentions petroglyphs in anecdotes, but he does not sketch a single one, or even describe any that he saw during his visit. Most telling is when he stands before the huge boulders and examines them most carefully: "The surface of these boulders are in every part entire and unscratched by anything: wearing the moss everywhere unbroken, except where I applied the hammer, to obtain some small specimens, which I shall bring away with me" (1859, 702). Yet, he says nothing of the petroglyphs that he must have stood on when he saw these great rocks. A sketch that a visitor to the site made in 1889, entitled *Native American contemplating the Three Maidens petroglyphs*, includes several of these small slabs, still in place and covered with glyphs (Scott et al. 2006, 235).

Furthermore, when he left Pipestone, Catlin asked his guide to detour on the trail back when they passed other interesting sites along the Couteau. From this it seems that no-one in this 19th century mix of tribal people and fur traders was aware of the thousands of petroglyphs carved on the rocks a short ride to the east.

Ancient travels

The overwhelming significance of Pipestone provides a reasonable explanation for the presence of Ojibwa rock art there, especially during the 18th and early 19th century, when a traveler could follow fur trade routes by canoe as Catlin did to get to the site. This still does not explain the presence of Ojibwa at Red Rock Ridge, however, and why it was so heavily used, before the 19th century.

As we suggest, the reason may not relate to ideas at all – but rather to the simple circumstances of travel across the grasslands and streams of southern Minnesota over the centuries. The journey by canoe and horseback that Catlin describes from Lake Michigan is the most efficient way to reach central and southern Minnesota from the north. However, as the petroglyphs on Red

Rock Ridge seem to relate to earlier times, it may be that these travelers used canoes or dugouts and then navigated the final overland stretch on foot.

One might assume that the fur trade route along the Cottonwood followed a more ancient route to the quarries. It was, however, an historical accident. LaFramboise wanted to exploit the trade in muskrat pelts, so he set up his trading post miles inland among the many small lakes on the Couteau and used oxcarts to carry the pelts along the Cottonwood to the Minnesota River (see Timmerman 2009, 25). From this time on, it became the chosen route to Pipestone for explorers such as Catlin and Nicollet and virtually all other visitors, all of them unaware of the large petroglyph site less than two days' ride to the east.

Ironically, the LaFramboise post lasted only a few years as the price of muskrat pelts collapsed in 1837 and Henry Sibley, the regional manager of the American Fur Company, ordered LaFramboise to quit his post and move back to the Minnesota River – but the trail persisted (Forrest 1947, 9). Eventually, the government built a wagon road above the Cottonwood River from Ft. Ridgley at its mouth to Ft. Thompson on the Missouri River and, today, U.S. Highway 14 follows the Minnesota part of the route from New Ulm on the Minnesota River to Ft. Benton, a few miles north of Pipestone.

Had LaFramboise not made this decision and set a path that others would inevitably follow, the Cottonwood River might not have appeared as the most efficient route towards the southwest, especially coming from the north and east through the Mississippi drainage, and especially by canoe and on foot. Rivers are extended landmarks that do not require complex mental maps or landscape narratives to follow, only the knowledge that the flow of a particular stream inevitably leads to its source or debouch. The Cottonwood is navigable for about half its length, but it swings towards the northwest, and so a direct journey towards the quarries had to be overland, with no stream course to use as a guide through the drift plains, and without the hospitable LaFramboise leading the way to the quarries.

There are, however, two optional routes. A few miles downstream on the Minnesota from the Cottonwood is a much smaller stream, with no large meanders or large tributaries: the Little Cottonwood. Rather than flowing from a westerly direction, it drains into the Minnesota from headwaters at the eastern end of Red Rock Ridge. Lothson thought that a migratory game trail active during the Archaic along its course would bring the hunters to the ridge (Lothson 1976). The other is the north branch of the Watonwan River, an east–west tributary of the Blue Earth River, which rises only 3 miles from the Ridge. The significance of this route is that the Blue Earth flows from the south into the Minnesota right where it swings north to its confluence with the Mississippi, thus creating a path directly to the Jeffers site.

The south fork of the Watonwan had some considerable significance to Joseph Nicollet. As a member of the United States Army Corps of Topographical Engineers, he was interested in the improvement of transportation across this region. In the report of his regional survey, he noted that

American Fur company traders wintered in the headwaters of the Des Moines and they knew of an alternative route to the fur trade post at the Traverse de Sioux on the Minnesota River (1841, 24).

> On one occasion, Mr. Joseph Laframboise, failing in his means of transportation by land, had a large canoe built, which he loaded with his peltries, took water upon the Tchan-shetcha, descended the Watonwon and the Mankato [Blue Earth] and arrived safely at the St. Peter's station.

Nicollet marked this one-and-a-half-mile portage between the Tchan-shetcha [Fish Lake] on his map. He imagined that one day a canal might be cut over this short portage, which would connect the Des Moines and Watonwan Rivers and create a continuous water route for future shipping that would cross southern Minnesota, run diagonally through Iowa and then turn north up the Mississippi and back to the Watonwan.

> I mean only to indicate at present, however, what will at a future day form an obvious feature in the system of internal improvement of these regions, so new in the geography of the United States, by which this extensive and beautiful territory might be rendered circumnavigable.
> (1841, 24)

Had he considered the north fork of the Watonwan and traveled there, he might have discovered the petroglyphs on Red Rock Ridge himself.

The awareness of fur traders of a route by portage and canoe that cut across the divide between the Mississippi and Missouri in southern Minnesota, one they no doubt learned from the indigenous inhabitants, may help resolve two mysteries: why the Jeffers site was so heavily used, probably over thousands of years, and why it was overlooked after the introduction of the horse and the fur trade. While travelers who reached Red Rock Ridge from the north or east still had an overland trek west, the long ridge at Jeffers points like a compass towards the rising and setting suns.

It is very easy to forget in archaeology that ideas and cultural material only move through the actions of individual people in real time, who did not simply act out descriptions, such as hunting, gathering, traveling or trading, but rather walked from place to place over hours and days or, if they could, used dugouts, canoes or other craft. If an east–west route between the Missouri and Mississippi basins existed in southern Minnesota before the introduction of the horse in the 18th century, or others like it, it may have afforded a simpler passage for people carrying exotic materials from the west, such as obsidian from Idaho and Wyoming, Knife River Flint from North Dakota and catlinite from southwestern Minnesota for trade with Hopewell and Mississippian groups in the northeast.

This is difficult to resolve in the archaeology, especially if the artifacts arrived already made, because the route would not be dotted with artifacts and the spoils of their manufacture. In an analysis of catlinite platform pipes from the Tremper Mound, a large Hopewellian earthwork constructed about 2000 years ago on the Scioto River in Ohio, Emerson et al. (2005) suggest one route from Pipestone northeast into Wisconsin and then around the tip of Lake Michigan, and the other, a much longer route, from the quarries down the Missouri and up the Mississippi and the Ohio (2005, 207–208). Twenty years earlier, Clark (1984) had proposed that Knife River flint, a prized toolstone from North Dakota that appears in Hopewell sites, might have been traded overland between Minnesota and Wisconsin (1984, 174). Perhaps an overland route across the two great watersheds, with Red Rock Ridge as a vector, was a conduit for both.

Setting aside the complex archaeological record for the geography, the most prominent regional landmark near Pipestone, one easily identified, recalled and passed on to other travelers, is not Blood Run, but rather the rapids and waterfalls where the Big Sioux River runs over a ridge of quartzite, a few miles upstream. This area shares a similar pattern of settlement, but the villages, most of the burial mounds and all other traces of early settlement were wiped out in the development of the city of Sioux Falls, South Dakota.

The potential of this area becomes clear when assumptions about movements along large rivers are also set aside. When environmental affordances to cross-country movements are mapped without reference to the archaeological record, the rather small Big Sioux River provides a strategic advantage over other routes. From its main course it is possible to travel in three directions. A journey upstream to the headwaters leads with a short traverse into the Red River, which flows north into the boreal forest; downstream, it enters the Missouri where it turns to flow south, eventually joining the Mississippi in its run to the Gulf of Mexico. As for the east, it is possible that the cross-Minnesota route described here was a way for people traveling by canoe and on foot to venture relatively quickly between the Great Lakes and northeast and the central Great Plains.

Considering the petroglyphs at Red Rock Ridge and Pipestone this way may help with the curious differences in their histories. If Red Rock Ridge was along an ancient overland travel route, favoured because it cut east–west across the contrary flows of streams flowing southeast towards the Mississippi River, its polished rocks would have inevitably drawn a few artists among the many travelers over thousands of years. For all these outsiders, it was probably only a stop along the way. With the quarries and the great intersection of rivers at the falls of the Big Sioux and at Blood Run, however, it was quite likely that visitors and traders coming to and from the quarries had more important matters to attend to than making petroglyphs.

How rocks speak

A common problem in rock art research is treating the art as if it appeared on the rocks fully formed. Even in a relational model, which imagines individuals and draws from ideas about the perception of landscape (e.g., Ingold 2000), the relation of the individual at the site may be with an absent, abstracted society.

The rocks speak by where they stand in relation to the people who lived nearby, whose situations, attitudes and desires influenced a handful of individuals to travel there to paint or carve, as part of the relentless movements in a hunting and gathering life. It does not diminish rock art in some circumstances to be incidental. Seeing it in the flows of ordinary life provides a releasement from governing attitudes and ideologies of all kinds, which control who and what can be said or believed, and what forms of study are even possible. It is also practical. Individual people are much closer to the circumstances that artists in the past faced than they are to their ideas. When the only alternatives in rock art research seem to be description and discourse, rock art researchers may find new fields of possibility by exploring the commensurate interactions of the artists in the surroundings shared at the sites. In place of a search for meaning, the challenge is to find evidence of the situations they faced.

References

Alex, L.M. 2000. *Iowa's Archaeological Past*. Iowa City: University of Iowa.

Anonymous. 1839. Catlinite or Indian Pipe Stone. *American Journal of Science* 35:388.

Arsenault, Daniel. 2013. The Aesthetic Power of Ancient Dorset Images at Qajartalik, A Unique Petroglyph Site In The Canadian Arctic. *Boletín Del Museo Chileno De Arte Precolombino* 18(2):19–32.

Babcock, Willoughby M. 1939. Louis Provençalle, Fur Trader. *Minnesota History* 20:259–268.

Birk, Douglas A. 1974. *An Archaeological Survey of the Painted Rock Cliff Site, Crooked Lake, Lake County, Minnesota*. Ms. St. Paul: Minnesota Historical Society.

Blakeslee, Donald J. 1981. The Origin and Spread of the Calumet Ceremony. *American Antiquity* 46:759–768.

Boszhardt, Robert R., and James N. Gundersen. 2003. X-Ray Powder Diffraction Analysis of Early and Middle Woodland Red Pipes from Wisconsin. *Midcontinental Journal of Archaeology* 28:33–48.

Bray, Edmund C. 1977. *Billions of Years in Minnesota: the Geological Story of the State*. St. Paul: Science Museum of Minnesota.

Brown, Ian W. 2006. The Calumet Ceremony in the Southeast as Observed Archaeologically. In *Powhatan's Mantle: Indians in the Colonial Southeast, Revised and Expanded Edition*. Edited by Gregory A. Waselkov, Peter H. Wood and Tom Hatley, 371–419. Lincoln: University of Nebraska Press.

Buhta, Austin A., Scott F. Anfinson, Eric C. Grimm and L. Adrien Hannus. 2017. *Minnesota's Archaic Tradition: an Archeological and Paleoenvironmental Overview and Assessment*, Archaeological Contract Series 292. Sioux Falls: Augustana University.

Callahan, Kevin L. 2001. *The Jeffers Petroglyphs: Native American Rock Art on the Midwestern Plains*. St Paul: Prairie Smoke Press.

Catlin, George. 1859. *Letters and Notes on the Manners, Customs, and Condition of the North American Indians. Written During Eight Years' Travel Amongst the Wildest Tribes of Indians in North America*. Two Volumes in One. Philadelphia: J.W. Bradley.

Catlin, George. 1899. *Life among the Indians*. London: Gall and Inglis.

Clark, Frances. 1984. Knife River Flint and Interregional Exchange. *Midcontinental Journal of Archaeology* 9:173–198.

Colson, Alicia. 2011a. Do Not Make Snap Decisions About What You are Seeing: How Digital Analysis of the Images from the Canadian Shield Highlights the Difficulties in Classifying Shapes. *International Journal of Humanities and Arts* 5:59–81.

Colson, Alicia. 2011b. What Do these Symbols Mean? A Critical Review of the Images Found on the Rocks of the Canadian Shield with Specific Reference to the Pictographs of the Lake of the Woods. *Revista de Arqueología Americana* 25:1–87.

Coues, Elliot, ed. 1897. *The Manuscript Journals of Alexander Henry, Fur Trader of the Northwest Company, and of David Thompson, Official Geographer and Explorer of the Same Company, 1799–1814. In Three Volumes. Vol. I, The Red River of the North*. New York: Francis P. Harper.

Delafield, Joseph. [1823] 1943. *The Unfortified Boundary. A Diary of the first survey of the Canadian Boundary Line from St. Regis to the Lake of the Woods*, edited by Robert McElroy and Thomas Riggs. New York: Privately printed.

Dewdney, Selwyn, H. 1975. *Sacred Scrolls of the Southern Ojibway*. Toronto: University of Toronto Press.

Dewdney, Selwyn, and Kenneth Kidd. 1962. *Indian Rock Paintings of the Great Lakes Region*. Toronto: University of Toronto Press.

Dewdney, Selwyn, and Kenneth Kidd. 1967. *Indian Rock Paintings of the Great Lakes Region*. Second Edition. Toronto: University of Toronto Press.

Dyke, Arthur S. 2004. An Outline of North American Deglaciation with Emphasis on Central and Northern Canada. In *Quaternary Glaciations – Extent and Chronology Part II: North America*, edited by J. Ehlers and P.L. Gibbard. Volume 2, Part B, 373–424. Amsterdam: Elsevier.

Eid, Leroy V. 1979. The Ojibwa–Iroquois War: The War the Five Nations Did Not Win. *Ethnohistory* 26:297–324.

Emerson, Thomas E., Kenneth B. Farnsworth, Sarah U. Wisseman and Randall E. Hughes. 2013. The Allure of the Exotic: Reexamining the Use of Local and Distant Pipestone Quarries in Ohio Hopewell Pipe Caches. *American Antiquity* 78:48–67.

Emerson, Thomas E. and Randall E. Hughes. 2001. De-Mything the Cahokia Catlinite Trade. *Plains Anthropologist* 46:146–161.

Emerson, Thomas E., Randall E. Hughes, Kenneth Blaine Farnsworth, Sarah U. Wisseman and Mary R. Hynes. 2005. Tremper Mound, Hopewell Catlinite, and PIMA Technology. *Midcontinental Journal of Archaeology* 30(2):189–216.

Fishel, Richard L., Sarah U. Wisseman, Randall E. Hughes and Thomas E. Emerson. 2010. Sourcing Red Pipestone Artifacts from Oneota Villages in the Little Sioux Valley of Northwest Iowa. *Midcontinental Journal of Archaeology* 35:167–198.

Forrest, Robert B. 1947. *A History of Western Murray County from 1688 to December 1946 and of Leeds Township: A Typical Prairie Township*. Lake Wilson, Minnesota: Published by the Author. http://files.usgwarchives.net/mn/murray/history/western/murrtoc.txt.

Giardina, Irene. 2008. Collective Behavior in Animal Groups: Theoretical Models and Empirical studies, *HFSP Journal* 2:205–219. doi:10.2976/1.2961038.

Gundersen, James N. 1991. *The Mineralogical Characterization of Catlinite from its Sole Provenance, Pipestone National Monument, Minnesota*. Research/Resources Management Report MWR 17. Omaha: Midwest Regional Office, National Park Service, U.S. Department of the Interior.

Hämäläinen, Pekka. 2019. *Lakota America: A New History of Indigenous Power*. New Haven: Yale University Press.

Hargrave, Joseph James. 1871. *Red River*. Montreal: John Lovell.

Henning, Dale R. 2007. Continuity and Change in the Eastern Plains, A.D. 800–1700: An Examination of Exchange Patterns. In *Plains Village Archaeology: Bison-Hunting Farmers in the Central and Northern Plains*, edited by S.A. Ahler and M. Kay, pp. 67–82. Salt Lake City: University of Utah.

Henning, Dale R., and Shirley J. Schermer. 2004. Blood Run Archaeological Investigations. *Plains Anthropologist* 49:399–434.

Henning, Dale R., and Thomas D. Thiessen. 2004. Regional Prehistory. *Plains Anthropologist* 49:381–398.

Holt, Edwin B. 1915. *The Freudian Wish and its Place in Ethics*. New York: Henry Holt and Company.

Ingold, Tim. 2000. *The Perception of the Environment: Essays on Livelihood, Dwelling and Skill*. London: Routledge.

Leverett, Frank, and Frederick W. Sarde. 1932. *Quaternary Geology of Minnesota and Parts of Adjacent States*. Washington: United States Government Printing Office.

Lothson, Gordon A. 1976. *The Jeffers Petroglyphs Site: a Survey and Analysis of the Carvings*. St. Paul: Minnesota Historical Society.

Mackenzie, Alexander. 1801. *Voyages from Montreal through the Continent of North America, to the Frozen and Pacific Oceans: In the Years 1789 and 1793. With a Preliminary Account of the Rise, Progress, and Present State of the Fur Trade of that Country*. London: R. Noble, Old-Bailey.

Molyneaux, Brian L. 1977. *Formalism and contextualism: An historiography of rock art research in the new world*. M.A. Thesis, Art and Archaeology, Department of Anthropology, Trent University, Peterborough, Ontario.

Molyneaux, Brian L. 1983. *The Study of Prehistoric Sacred Places: Evidence from Lower Manitou Lake*. Archaeology Paper 2. Toronto: Royal Ontario Museum.

Molyneaux, Brian L. 1986. Marten Heaven: a Painted Rock in Quetico. *Royal Ontario Museum, Archaeological Newsletter, Series* II 12:1–4.

Molyneaux, Brian L. 1987. The Lake of the Painted Cave. *Archaeology* 40 (4):18–25.

Molyneaux, Brian L. 1998. Recent Clovis Point Discovery. *Iowa Archeology News* 4:5.

Molyneaux, Brian L. 2002. Exploring the Landscapes of Long-distance Exchange: Evidence from Obsidian Cliffs and Devils Tower, Wyoming. In *Geochemical Evidence for Long-Distance Exchange*, edited by M. Glasscock, 133–151. Westport: Bergin & Garvey.

Nicollet, N. 1841. *Report Intended to Illustrate a Map of the Hydrographical Basin of the Upper Mississippi River*. Washington: Blair and Rives.

Nute, Grace Lee. 1948. The Picture Rock of Crooked Lake. *Minnesota History* 29:130–136.

Qajartalik, A Unique Petroglyph Site in the Canadian Arctic. *Boletín Del Museo Chileno De Arte Precolombino* 18:19–32.

Sawdo, Phillip T. 2004. *The Trapper*. Victoria: Trafford Publishing.

Scott, Douglas D., Thomas D. Thiessen, Jeffrey J. Richner and Scott Stadler. 2006. *An Archeological Inventory and Overview of Pipestone National Monument, Minnesota*. Midwest Archeological Center, Occasional Studies in Anthropology No. 34. Lincoln: National Park Service, Midwest.

Swanson, Evadene Burris. 1940. *The Use and Conservation of Minnesota Wildlife 1850–1900*. PhD Thesis, University of Minnesota, Minneapolis.

Taliaferro, Lawrence. 1836. *Taliaferro Journals, Daily Transactions and other Matters on Indian Affairs, St. Peters Agency*, 10th Volume, 152–153. St. Paul: Minnesota Historical Society.

Tapper, Bryn, Oscar Moro-Abadía and Dagmara Zawadzka. 2020. Representation and Meaning in Rock Art: the Case of Algonquian Rock Images. *World Archaeology* 52:449–462. doi:10.1080/00438243.2021.1897660.

Timmerman, Janet. 2009. *Joseph Laframboise: A Factor of Treaties, Trade, and Culture.* M.A. Thesis. Manhattan, Kansas: Department of History, College of Arts and Sciences, Kansas State University.

Tyrrell, J.B., ed. 1916. *David Thompson's Narrative of His Explorations in Western America 1784–1812*. Toronto: The Champlain Society.

Willey, Gordon R., and Philip Phillips. 1957. *Method and Theory in American Archaeology.* Chicago: University of Chicago Press.

Zawadzka, Dagmara. 2013. Beyond the Sacred: Temagami Area Rock Art and Indigenous Routes. *Ontario Archaeology* 93:59–199.

6 In search of a situation

Ideological space

There is an outcrop of soft, smooth black slate along the Medway River in south-central Nova Scotia that is now under the waters of a small hydroelectric dam. In the 19th century and probably much earlier, Mi'kmaw inhabitants of the area carved petroglyphs there.

This chapter ponders a specific incident, where one carver has taken two existing petroglyphs and reworked them into new forms, with new meanings and significance. While one could talk about how they mark the transition between conceptual worlds (e.g., Molyneaux 1989), the goal here is the consider the situation itself, which ended with the second carver's transformative act.

It is not simple to begin again from a site, however – and not because of historical or cultural relativism. Even in the moonscape of the drained and sodden McGowan Lake, the slates and surrounding environment still attend every act. There are two related issues. The first is the treatment of the site as an instance in a larger set of images and ideas, as if the artist's role was service to cultural histories or processes. And the second is the application of the received wisdom back on the rock, which transforms the image into a representation and its surroundings into a conceptual environment – a matrix or landscape. In order to clear a path among these ideas to look at the direct relations of the carvers and surroundings, we need to account once again for an analytical return to physical space from ideological space and to environment from landscape.

Ideas about creating space for ideology in the environment converged in the idea of landscape late in the 20th century. In rock art, it began as a practical way to apply evidence of beliefs and activity associated with specific places in a local environment, making use of oral traditions, ethnohistory and archaeology, to create a map of site potential that would aid in archaeological survey (Molyneaux 1983). As the idea of landscape developed in archaeology, however, interest shifted from practical, map-like distributions to the dynamic social processes responsible for these patterns. This approach was fundamentally different, as it replaced direct material relations with intellectual models and representations.

DOI: 10.4324/9781351053105-6

In order to deal with the lack of fit between general ideology and the specific practices of people in the past, some analysts have turned to phenomenology and, specifically, to the phenomenology of landscape. This approach uses individual experience in specific places and settings, taken from first-person descriptions and reveries, as clues to how people might have perceived these environments in the past. Within a landscape paradigm, it seems reasonable to imagine inhabited archaeological space as conceptually and culturally relative and, indeed, in these terms, the space of the Paleolithic is not the same as space in the 20th century as Tilley once suggested (e.g., Tilley 1994).

This is the space of ideas, however, not where people live. Landscape phenomenology acknowledges the persistence of environment and the commensurability of human actions and perceptions over time, yet the concept of landscape encloses this behaviour in a cultural totality – as if the environment only exists to the extent that philosophy allows, and only in ways defined within its self-involved discourse. As interpreters ponder the vicissitudes of subjective experience within a larger world of artifacts, features and relations in a locality, the landscape becomes a shelter for the imagination. This model is well expressed in Bachelard's evocative idea of a house (1994, 47).

> A house that has been experienced is not an inert box. Inhabited space transcends geometrical space.

Thus, landscapes are not only conceptual enclosures, they are domains of infinite discourse, as Tilley and Cameron-Daum (2017) demonstrate, and this is a problem for archaeology. While the concept of landscape tacitly accepts that it is impossible to escape one's mind, it does not follow that some collective force or turn of phrase unites and governs social activity. If, as Tilley and Cameron-Daum observe, "The thinking, subjective mind emerges in relation to the landscape and ends in its perception" (2017, 7), humanity is forever trapped in a cycle of self-discovery and social affirmation. What is missing in such ideological approaches is any sense that a world exists entirely outside our self-preoccupations. Part of this territory is the uncharted, uncalculated sensory interactions that we have in common with other creatures –and another is the ordinary life carried out on the ground, the one that individual people in their groups and societies deal with daily, that is simply external to all the metaphors of inhabitation conceived in discourse. As Sartre observed about individual experience:

> if we know that we are directors of being, we also know that we are not its producers. If we turn away from this landscape, it will sink back into its dark permanence. At least, it will sink back; there is no one mad enough to think that it is going to be annihilated.
>
> (Sartre 1948, 26)

The problem of landscape is thus not a problem of description, but of scale and the choice of moments in time. Archaeology began with curiosity about things on the literal ground, not the thingness of ideas. This was nothing special; it is the behaviour of anyone who, walking by an eroding bank or along the seashore or a riverbank, spies something odd, picks it up and wonders, all without instructions or other institutional support. While the thing possessed inevitably becomes a product in an economy of speech about it, within a wider intellectual industry, this is its latter-day life. The root of archaeology [ἀρχαιο-] in the Greek has senses of ancient, primitive, beginning (Simpson and Weiner 2009). This does not refer to the beginning of discourse, which is the secondary processing of the object – it refers to the primacy of discovery and encounter, what took place *before* the object, the circumstances that preceded and precipitated the act.

Searching for the evidence of situations before the object and its inevitable narrative forces the researcher to think again about the site of discovery. Here, as in rock art, or other places where some traces exist in their original settings, the material remains of the past persist like the rest of brute nature as part of the natural environment.

Across the golden divide and, therefore, unencumbered by phenomenology, humans share the invariants and affordances of shelters in the physical environment with many other creatures. In this way, the spaces of Pleistocene caves, late capitalist buildings and ordinary abodes are much the same outside their cultural trappings. They indifferently provide material and social shelter, accommodating humans, bats, rodents, insects and other creatures – and walls for paintings.

The problem with concepts of environment is a simple bias. A shelter is what is, but the idea of a shelter is always an opinion. And while we may be stuck with metaphors, there is ground in each, and it is a matter of being careful about the relations implied in them – that is, whether the values we attribute to them are of the earth, rather than society. Consider, for example, the rich metaphorical traditions associated with the common oppositions of light and darkness, settled and wild, and sacred and profane.

Humans, as diurnal creatures, lack control in darkness, and so it has a natural anarchy that offers freedom from constraint and the potential for danger and threats to social order. In a city, street lighting pushes back the darkness and, it is generally assumed, reduces crime, just as in the popular imagination, the light of campfires kept back the creatures of the night. However, the reality is somewhat different. In forensic geography, studies of the relation of street lighting and crime have mixed results. Brantingham and Brantingham note that "places marked by darkness and isolation are generally feared as likely crime sites, but (with a few exceptions) tend to be relatively low-frequency crime locations" and "the introduction of higher levels of street lighting into high-fear locations" tends to have little impact on levels of crime (1995, 6). Most recently, O'Connell (2017) came to a similar conclusion in a study in Houston, Texas. And in the UK, Perkins, Steinbach et al. (2015) reported that in a study about plans to limit street lighting as a cost-saving device, there was "little evidence of harmful effects of switch-off, part-night lighting, dimming or changes to white light/LEDs on road collisions or crime in England and Wales" (2015, 17). There is a simple environmental explanation for these inconclusive studies – crime takes place in the light because perpetrators also need to see. As Eck (1997, 22) concludes, "Who the lighting serves is unclear".

Popular assumptions that absence of light is a problem reveal a general inattention to direct sensory experience. Lights can get in the way. Imagine walking at night along a fully lighted street, or following the tunnel of light from a flashlight or torch; the light attracts attention and the brightness makes it impossible to see who or what might be watching in the wall of darkness beyond. Darkness is a question of attitude that one can adjust to. People who live without artificial light spend a significant part of their active lives in twilight or darkness – in the Arctic, months this way – as everyone used to, and they move around at night by moonlight or bring in other senses, like sound and touch, as non-sighted people do throughout their lives. Thus, the reality of darkness across time and culture is not quite the psychological and social barrier we make it out to be (see Bordin 2015).

Another concept that has emerged in urban settings is wilderness, the unfamiliar outskirts of settlements. Even the wildest of these outlands is not so to its inhabitants. The home range of a hunting and gathering society is as familiar as a neighborhood in an ostensibly tamer and more predictable world.

Most subtly, and extremely influential in studies of prehistoric art, the idea of darkness and wilderness are essential aspects of religions, as they are brought into the articulation of sacred space. The sacred is commonly viewed as a refuge or oasis in an uncontrolled world that is primal, liminal, benevolent or dangerous. In the Bible, God created the world out of darkness and land outside his protection was "waste howling wilderness" (Deuteronomy 32,10 KJV).

After thousands of years of settled life and the increasing technical control over interactions with the rest of nature, these metaphors have become entrenched in western epistemologies and, therefore, influence attitudes within research about the past. In a recent study of the use of artificial light in Paleolithic caves (Medina-Alcaide, Garate et al. 2021), for example, the authors resist speculation by evaluating the nature and capacities of the lighting devices actually used in painted caves and, yet, they still assume that Paleolithic people conquered the darkness and conducted activities that are "closely linked to the origins of human symbolic and artistic behavior".

The people who live in outlands, on the earth at the bottom of the ladders and chains of being, or in ancient times, where they frequented the deep passages in caves, are not simply "others", the exceptions to the rule or, alternatively, those that should occupy the center. They are all routinely denied the capacity to be ordinary, to exist as they choose outside discourse about them. For example, indigenous hunters and gatherers tend to have relatively egalitarian social organizations and, hence, more equal distributions of power and wealth than sedentary societies. If such traits are regarded as exceptional, however, they are denied the capacity to have some of the social mores and behavioural traits commonly identified with sedentism, such as concepts of self, selfishness, pragmatic self-interest, personal property and individualism – even though hunting and gathering require considerable self-reliance on the part of individuals and families within these larger social groups.

In rock art, consider the burden of the exceptional in the search for meaning – and how it might distort what the images might mean *to the artists and their communities.* It cannot be assumed that the psychological or emotional attitude about darkness in sedentary societies was even remotely shared by Pleistocene foragers, who often lived in caves and shelters, or that hunters and gatherers had a sense of wilderness outside their familiar grounds.

Recalling the wakan of the Sioux and the Chinese Tao, the idea of the sacred may be similarly distorted, especially in societies where religiosity is individual, rather than collective and canonical, and the models for worship are in the natural surroundings. To visualize this problem, imagine sacred power like a chain of being, from the more or less secular ground to the most intensely spiritual world in the unreachable sky. This makes practical sense in any society that believes that spirits occupy the sky. The highest points of land are very often the chosen objects of veneration because the higher worshippers can climb, the closer they are to the heavens. An important aspect of Adam's Peak to Muslims and Christians in medieval times was that it seemed to be the highest mountain in the world, and so, the most likely place for Adam to have come down to earth.

In the Black Hills, just beyond the northern edge of the uplift, a large, isolated mountain sits strangely on a featureless plain. The Lakota call it Mato Paha (Bear Butte), because at some angles, it resembles a bear at rest. This great hill is central to the spiritual identities of the Lakota and Cheyenne nations, and many tribal members make pilgrimages there, climbing up its steep flanks to the summit for prayers and rituals. It is also a South Dakota state park, and a paved road that runs up to a visitor center midway up the hill provides access to its upper slopes. Near this facility, at a bend in the road, prayer ribbons and tobacco packets are sometimes visible in trees and bushes close to the pavement. This spot is well below the summit – but for elders who can go no further, and for others who cannot or choose not to climb any higher, this place must still be appropriate for their prayers and offerings, and we should assume that the efficacy of their gestures is undiminished.

Such brief and private acts by ordinary people, carried out in a public place over a few minutes in a particular day, expose the inadequacy of general categories and stereotypes as ways to characterize ideas and practices associated with belief. Such representations may descend like low cloud over societies and their intellectuals and limit more detailed, critical explorations of lived life, past and present.

The incompatibility among even fundamental ideas about the experienced world justifies the caution here with notions of representation. In the 21st century, interpreters should not simply assume that received narratives about belief from any source fit the past they are attempting to understand. Overviews may reveal processes and events that resolve at wider scales, but staying close to the ground provides a better sense of lived life in the past.

Rock carvings at McGowan Lake

Setting and surroundings

The Nova Scotia peninsula is part of the Appalachian Mountains, an ancient and greatly eroded chain that extends from Newfoundland to Alabama. In a topographical cross-section, the land rises gently from the outer Atlantic coastal fringe to the modest heights of the mountains and then drops more steeply near the inner coast to the Bay of Fundy, which separates the peninsula from the mainland. In the southwestern interior, lakes that formed in a relatively level upland during the glacial retreat (ca. 13,000 BP) flow in several large rivers southeastwards into the Atlantic. This entire region is Mi'kmawa'kik, a homeland of the Mi'kmaq, an Algonkian people. As the peninsula is so narrow, only 70 miles from coast to coast, seasonal movements between the interior and sea have shaped Mi'kmaw life and, because of the ease of navigation, afforded diverse interactions with other Tribes and with Europeans, who crossed the Atlantic to exploit these resources.

The first European encounters were probably with Vikings who settled briefly in Newfoundland during the 11th century and ventured southwards across the Gulf of St. Laurence to Cape Breton and some distance further along the coast (Ingstad and Ingstad 2001). Basque, Breton and Norman fisherman and whale hunters from coastal France, Portugal and Spain followed, fishing the banks off Newfoundland and southwestern Nova Scotia and processing their catches along the coasts (LeHuenen 1984). Interactions were frequent enough that a pidgin seems to have developed among the Mi'kmaq and other coastal Algonkian tribes and these Europeans, which persisted for about 100 years from about 1540 to the mid-17th century (Bakker 1989).

The French established their first colonial outpost on the peninsula in 1605 on the shore of the Annapolis Basin, a sheltered inlet along the Bay of Fundy, at a place they named Port Royal (Thwaites 1896a). This was not simply a staging ground for the maritime economy; there was a compelling need to establish political and social control over the lands that France had claimed – of course, without asking the Mi'kmaq.

By this time, the Tribe had already mastered the use of their resources and technology. This was evident in Mi'kmaw seafaring skills. They hunted dolphins and whales using sea-going birchbark canoes, and quickly adapted to shallops and other European vessels. Marc Lescarbot, a barrister who Jean de Biencourt, Seigneur de Poutrincourt, enlisted to help him look after the nascent settlement of Port Royal, wrote in his journal about an encounter with two shallops on the open sea during their voyage in 1606. One was manned by Mi'kmaq and the other, by "Frenchmen of Saint Malo, which made their fishing at the Port of Campseau". The Mi'kmaq, whose craft had a stag painted on its sails, "were more diligent, for they arrived first" (Lescarbot 1928, 84).

Over the next two centuries, the territory was the focus of colonial rivalries between France and Great Britain over a Mi'kmaq homeland that was subject to persistent exploitation of its population and natural resources. After French colonization, Mi'kmaw people developed close trading and social relationships with the incomers, and most took on beliefs and traditions of the Catholic religion. When the British assumed control in 1763, the Tribe entered a long period of much more difficult relations and, before the end of the century, after the resettlement of thousands of British loyalists on the peninsula who fled New England after the American Revolution, they lost their military power and control of the land. Eventually, after 1867 and the formation of Canada, the Tribe was pushed into small reserves scattered across the province.

Because incoming settlement over this period was based primarily on maritime economy and agriculture, rather than the fur trade (Murray 1938), colonial expansion was much slower in the forests, lakes and river headwaters in the interior, which were the seasonal hunting grounds of the Mi'kmaq. The shoreline surrounding the largest of these lakes, Kejimkujik, has a remarkable feature. The glacial ice exposed a terrane of soft slate, and where it outcrops along shores, waves and ice have worn it smooth over thousands of years. The surface is so soft that it can be scraped, scratched or carved quickly with a bone or metal tool, or even a sharp stone – and the marks will survive until they too are removed by the impacts of waves and ice. As a result, everywhere these slates are exposed, they have petroglyphs, often in complex masses of overlaid images, with only the most deeply carved visible in the daylight.

In 1888, a local postmaster named George Creed found and recorded petroglyphs on similar slate outcrops along the Medway River, about 10 miles east of Kejimkujik Lake (Mallery 1893). Soon after, loggers built a timber dam, flooding the site, and this was followed in the 1940s by a hydroelectric dam, forming the McGowan Lake reservoir. In 1983, the dam was opened for repairs, the water receded and exposed the rocks, and all visible glyphs were recorded (Molyneaux 1989).

In one spot, the glyphs form a complex palimpsest with a bewildering range of subjects associated with life over the centuries, from typical animals and humans to an encampment of canvas tents and a finely sketched two-storey building resembling a Georgian townhouse. The haphazard orientations and numerous superimpositions suggest that most carvers did not see any earlier glyphs beneath their work.

The carvings

We are here to look at images by two artists whose work is separated by an unknown period of time. The first carved pictures of moose in two separate groups, several feet apart on the rock. In one, the animals are arranged in three registers: at the top, a bull moose, with its distinctive palmated antlers; in the middle, a bull, followed by a cow (or yearling) and another bull; and at the bottom, two bulls opposing one another. Detailed recording also revealed part sketches of at least four other moose in the upper registers by the same hand. They are randomly placed, as if they were failed attempts or preliminary sketches. The second group nearby depicts two bull moose, one following the other.

The carvings cannot be dated, but there are two sailing vessels in the palimpsest that resemble galleons or merchantmen used by traders between the 16th and 18th centuries. One vessel has two masts with gaff-rigged sails and a headsail, a high afterdeck and lower waist deck, while a similar craft nearby has exaggerated figures in what appear to be greatcoats. Under ordinary circumstances, the erosion of the slate would probably rule out glyphs as early as the 16th century, but this location, unlike Kejimkujik, was not exposed to waves and wind-driven ice before the dam, and this, along with the inundation, might have preserved them longer.

Sometime later, a second carver knelt in the same spot and sketched what appears to be a family group over the moose and one of the sailing vessels. A man, woman and child stand next to a woman riding sidesaddle on a horse, all of them facing the viewer; below them, another man appears in profile, along with two more women. By their dress, the people appear to be Mi'kmaw of the 19th century: the men wear coats gathered at the waist and beaver hats, and the women and a child wear long skirts and bonnets. The standing woman is the largest figure, her skirt and apron have geometric patterns that would have been either painted or appliqued, and she wears a necklace.

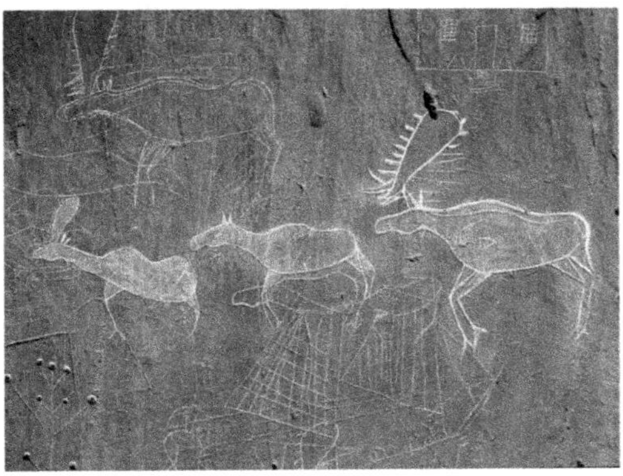

Figure 6.1 Moose Petroglyphs, McGowan Lake

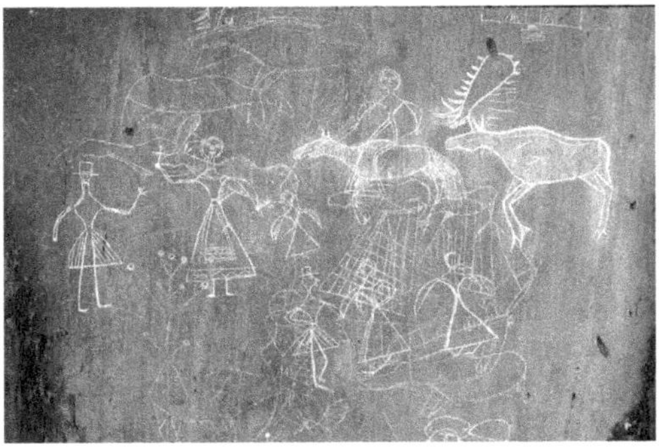

Figure 6.2 Family Group, Superimposed

The standing couple gestures towards anticipated viewers. The woman brandishes a cross in her right hand and a frond in her left, and the man beside her raises his arm in a gesture. The iconographic potential in this group is palpable, but it is the woman mounted on the horse that is the most telling. Instead of sketching a horse, the carver simply scratched a long, bushy tail over the haunches of the cow moose already there.

The same carver also created a scene out of the second group of moose, adding a Mi'kmaw man wearing a beaver hat and smoking a pipe and two dogs, one which attacks the moose from the front, and one from the rear. An inscription above was probably made by this second carver. The date is legible: November 1844.

A first impression of these petroglyphs is that they mark a conceptual shift from a preoccupation with animals to a more self-conscious depiction of humans under the pressures of acculturation (Molyneaux 1989). One should not assume from this, however, that it also shows the loss of cultural integrity. There is a profound difference between tradition and the living of it. Like all societies, the living society of the Mi'kmaq is not an ethnological or ideological construct. Over the generations, Mi'kmaw people either endured the

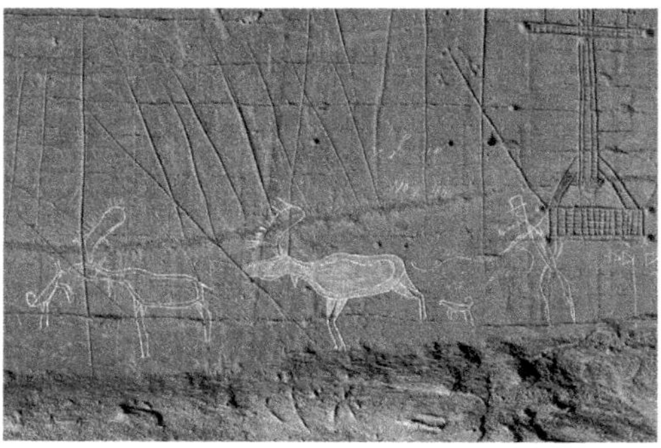

Figure 6.3 Moose with Hunter and Dogs, Superimposed in 1844

colonial struggle of the French and English over the eastern half of the continent directly or through its lasting social, ideological, economic and environmental effects. And like all other societies, they are different than the idealistic or stereotypical representations offered for them; they live in small, widely distributed communities whose inhabitants adapt to the local situations in which they find themselves (e.g., see Poliandri 2011). With such a diverse and complex background, there are no simple choices as to how to relate their art to the larger society, past or present.

Given this social complexity, the potential for further interpretation of the rock art on the Medway River may seem rather low, outside its broad cultural allusions. Yet, unlike the data of other aspects of cultural life, pictures are not actions observed, or speech translated and edited by others, but rather the direct, spontaneous expressions of individuals, marking brief episodes in their daily lives. Archaeology has a share of the study of endlessly created and interpreted material culture, and rock art research may also do so here, and study their latter-day life far from the rocks. If we look instead from the analogue images and ponder the situation of their creation, and follow the lead of the carvers and move outwards into the deeper surroundings, there is a chance that the images might record – not represent – evidence of past life in local Mi'kmaq communities in the Kejimkujik area, rather than being a memorial to it, preserved in symbols and traditions.

With these epistemological considerations in mind, we can now ponder what circumstances may have provoked the second Mi'kmaw carver to transform an iconic moose into a riding horse.

In search of a family

This portrayal of a family group is unique among the petroglyphs in the region, but there are other carvings from the 19th century that refer to actual people. A notable example is a moose carved at Fairy Bay on Kejimkujik Lake that is labeled "Jim Charles, his moose, Sept. 19, 1867".

To begin to understand why the carver did this, we can refer to events that directly affected the people in the region. A personalizing tendency may be seen in the social dynamics in colonial and European societies of the time. Among the Mi'kmaq and the other indigenous peoples who began to experience and adapt to the ways of Europeans, one of the most compelling inducements came from trade, first in the exchange of tributes and then the bartering of goods. In North America, the materials and technologies that incoming strangers possessed or offered – guns and other metals, wool and cotton goods by the yard, glass beads and other adornments, copper and iron pots, distilled spirits – were so novel and functionally advantageous that they were socially irruptive, stimulating the desire for material possessions among individuals and families and encouraging competition (e.g., Mead 1937; Yellen 1990). In North American rock art, James Keyser defines a "biographic style" in petroglyphs in the Great Plains, beginning after about 1750, that seem to be "records of personal war exploits" (1979, 44). The individualizing impetus here was the introduction of the horse in the region (see Ewers 1955). This new mode of transportation and form of wealth, combined with the push of tribes resident in eastern North America west as the Iroquois and Ojibwa became dominant, intensified competition over lands and resources among tribes settled around the periphery of the Great Plains.

The most devastating change, however, was a direct attack on collective values. Along with material things, epidemics, incoming traders, military forces and settlers, indigenous people had to deal with novel ideas about spirituality, brought by missionaries from France, and these ideas were reinforced by pictures. In Algonkian territories along the east coast and the eastern boreal forests, the Jesuits came with explorers and traders at the beginning of colonization, literally on a mission to force this conceptual change. Their incursion began on the eastern coastal peninsula that formed part of France's new colonial territory, Acadia, in 1611, when the Jesuit Fathers Pierre Biard and Ennemond Massé landed at the newly founded outpost called Port Royal, along the Bay of Fundy, with the mission of converting the Mi'kmaq to Christianity. This tiny outpost was just 34 miles (55 km) north of the petroglyphs on the Medway River and Kejimkujik Lake.

The occupation began in a climate of tension. France had just concluded a series of religious wars between Catholics and Protestants with the Edict of Nantes (1598), which provided rights for Protestants, and this created a new urgency in Catholics to promote and protect the faith. As with any colonial religious mission, the intent in the higher reaches of the Church and government was to eliminate indigenous people as a strong political and social entity. This was especially important on the east coast of North America, because in 1606, just before the Jesuits arrived, the British granted a charter giving two companies control of the lands and waters along the east coast south from present-day Maine (Bemiss 1957).

Through intermittent contacts since the late 15th century, the Mi'kmaq and French knew of each other's existence, but the incomers did not understand the language and had no measure for the kind of sophistication that the Tribe presented. This is when they turned to works of art. Months after establishing their mission in 1611, Father Biard sent a progress report to the Reverend Father Christopher Baltazar, Provincial of France, at Paris. In it, he recounts that while visiting a small post on the Saint John River, across the bay from Port Royal, he encountered five native converts and, he writes, "I took occasion to give them some pictures" (Thwaites 1896a, 163). Apparently, this was his habit when visiting a camp or village. He "prayed God in their presence, and showed them some pictures and tokens of our faith" (1896a, 225).

In one way this was simply pragmatic. Pope Gregory I had sanctioned the use of art in churches for the education of the illiterate masses in Europe a thousand years previously and this was strongly reaffirmed just a few years before the voyage at the final session of the Council of Trent (1545–1563), an ecumenical council that set out the ways and means to confirm and strengthen the dogma of the Church and counter Protestant iconoclasm. Almost at the end of this great council, bishops were specifically instructed to use pictures in religious education:

> And the bishops shall carefully teach this; that, by means of the histories of the mysteries of our Redemption, depicted by paintings or other representations, the people are instructed, and strengthened in remembering, and continually reflecting on the articles of faith …
>
> (Buckley 1851, 214)

There was more to such pictures, of course, than meets the eye. As they combined visual and ideological salience, they had effects independent of their assumed meanings and functions. At Port Royal in 1611, the catalyst was the choice of the Jesuits to carry the faith to the Mi'kmaq.

The Jesuits had an especially strong interest in visual representation – imaginative and physical. As a divinity student, the Society's founder, Ignatius of Loyola, wrote a small book of instructions, *Spiritual Exercises*, published between 1522 and 1524, to aid priests in their spiritual growth. His radical turn in the study of the Gospel was the emphasis on meditation, in which the priest would imagine participating in the events of the Gospel. He believed that a priest had to devote his whole body, senses and mind, that "a visualization of the scene or event, engaging all of the senses, would facilitate a better understanding of the life and teachings of Christ" (Fleming 2019, 188). While Ignatius emphasized the imagination, his few words on images, which he phrased as a rule in the *Spiritual Exercises*, were potent: "To praise the ornaments and the buildings of churches; likewise images, and to venerate them according to what they represent" (Mulan 1914, 206).

The Jesuits took this literally as they began to promote their version of the faith outside Europe in the late 16th century, carrying copies of important religious paintings and even altarpieces to their missions. Not surprisingly, inhabitants in these destinations were often mystified or confused. In 1581, a copy of the Virgin and Child after St. Luke, at the Jesuit residence in Kuangtung Provice was hidden because Chinese people mistook it for the Buddhist figure of Kuan-yin, who is sometimes drawn with a child in her arms (McCall 1948, 47).

In the Americas, however, the situation was different. The missions in China followed centuries of trade and other contacts, and so these ministrations were not irruptive – or effective. On the northeast coast of North America, however, hunters and gatherers had languages, beliefs and ways of life new to the Jesuits. The missionaries did recognize the difficulty ahead, but they did not see the greatest problem – their own inability to understand that the integrity of indigenous cultures, developed over thousands of years of adaptation, did not need correction.

As the gulf was so great, the Jesuits brought out the pictures, thinking they could educate and pacify tribal converts just as Ignatius had envisioned their own education. This was greatly aided by the publication in 1593 by the Jesuits of Antwerp of a book of illustrations of the Gospel stories, from the Annunciation to the Last Judgment. It was widely disseminated across Europe and in the missionary territories (Dekoninck 2017, 341), and probably brought to Acadia.

What Biard's hoped-for converts saw in the images may be assumed from an account at another Jesuit mission, teachings that included what Thwaites described as "the lurid pictures of the Judgment" (Thwaites 1896a, 22). In 1664, Father Henri Nouvel traveled with some Papinachois (an Innu band resident in northeastern Quebec and Labrador) along the north shore of the St. Lawrence River and up the Manicouagan River to the shores of its lake, where they erected a small chapel. Here he describes teaching a father and his son (Thwaites 1896b, 65–66).

> I was explaining to them the last judgment, letting them see, on a large picture where all was represented, what would be the happiness of those who believed in God, hoped in him, and loved and served him to the end; and, on the other hand, what would be the misery of those who did not believe in him and obey him: how the good Christians would be the companions of the Angels in Heaven, and the Infidels and bad Christians the companions of the Demons in the fires of Hell. At this point one of those good Catechumens interrupted me, exclaiming: "… My Father, in truth, we utterly hate the evil spirit. I beg you, let us not look at him any longer; let us always fix our gaze above. Oh, what pleasure we take in contemplating Heaven, and those who are happy there!" At the same time, perceiving that his eldest son, about twelve years of age, had his eyes fixed on the representation of Hell, he rebuked him: " … My son, thou hast no sense; look always upward."

Implicit in this enthusiastic report to a superior is the assumption that with sufficient education, converts would conjure up the correct mental pictures and, as with priests, make the spiritual connection. Instead, what they established probably stopped short of the Ignatian ideal. This was no doubt the first time that Algonkians had seen pictures of individual people. Unless it was "one of those good Catechumens" (a convert under instruction but before baptism), pictures hinted at heavenly rewards in the clouds, but graphically depicted a terrible, tangible fate in the fires of Hell.

As a result of this disconnect, missionary Christianity was also a more superficial religion of images and objects that carried a different kind of ideology forwards. It must be said, however, that its individualizing tendency would not have been out of place in ordinary circumstances in the relatively egalitarian societies of Algonkian peoples. Societies may manifest general degrees of individualism and collectivity (e.g., Mead 1937; Triandis et al. 1988), but in daily life, an individual routinely faces decisions that vary between individual and collective concerns, depending on the situation. Such personal flexibility is evident in the behavioural ecology of many indigenous societies, when family groups gathered or dispersed, depending on the season and the availability and access to food and other necessary resources. The individualism that the Jesuits introduced in pictures, however, was imposed, rather than emergent, and so had no practical ebb and flow of social responsibility. Beneath the pictures, lessons, orations and gift-giving was the implication that obedience would bring greater material rewards. This was not accidental. The commodification of spiritual desire hid the truth of the matter: the Jesuits were there to help control the existing population and facilitate the taking of land, the extraction of resources and the maintenance of power in the region.

Father Nouvel's letter in 1664 contains another anecdote that shows these contrary messages (in Thwaites 1898b, 69). At his mission, he encounters "a captain of the Oumamiois [another Innu band]", of whom he says "This good Catechumen, whom I baptized with his wife and four children, was never tired of speaking in praise of our mysteries". Spiritual rewards inevitably came with earthly ones:

> Sieur Amiot made him a present of a roll of tobacco, a sword, and some other things which they value, and I gave him two fine Pictures. In one of these the Mother of God was represented holding in her arms her Son Jesus; and in the other was the Savior of the world, holding a globe in one hand [the Salvator Mundi was a popular image among the Jesuits for use in their missions (e.g., McCall 1948); one wonders what they made of the globe]. At this he said wonderful things to us; he was going to exhibit these Pictures which I had presented to him to all the nations allied to his own, and he would make a tour of all the Villages which lie along the North Sea, to invite all the Inhabitants to prayer. He said that he would tell them in advance what I had taught him; that all the Captains of that country would taste the tobacco which Sieur Amiot had given him; and that the sword of which he had made him a present would speak in a loud voice in honor of the French.

Nouvel's effusive and blatantly political account, with its emphasis on the prospect of trade goods and the promise of even more rewards on earth, revealed the base colonial agenda. From their arrival, the Jesuits and other missionaries may have worked selflessly, but their mission lacked a sense of social and economic reciprocity that the Mi'kmaq might have expected from an intertribal relationship. Behind the early diplomacy and seemingly fair exchange of desired goods was a system of land appropriation, social pacification and resource extraction.

Returning to Mi'kmawa'kik, one might wonder why Mi'kmaw people entertained such promises, as they had vastly more experience in dealing with the traders and fishermen than interior Algonkian peoples. However, Murray (1938) estimates that by 1600, over a thousand ships visited the east coast every year, and the anticipation of gaining these wonderful new resources by bartering furs was no doubt high as soon as word spread along the coast. After initial contacts, however, this desire to trade might have been more of a necessity. By the 1600s, the Tribe may have lost as much as half its population through disease brought over with the fishermen and traders (Burley 1981).

As the centuries moved on from this portentous beginning in Mi'kmawa'-kik, Acadia and Nova Scotia, colonists increasingly occupied and exploited Mi'kmaq lands while maintaining a steady disregard for Mi'kmaq culture as a fundamentally valid, well-adjusted and well-integrated way of life. The Tribe continued to assert its rights as given in several "Peace and Friendship" treaties signed with the British in the 18th century, which gave them their traditional freedom to move at will through their territories and, in the wording of the 1752 agreement, maintain "free liberty of Hunting & Fishing as usual" (Akins 1869, 684). Such agreements were obviously subversive and fragile, however, and after the American Revolution in 1776, a large influx of Loyalists was resettled without any notice of Mi'kmaq treaties (Chute and Speck 1999).

As the colonial population increased and the logging industry developed, the Mi'kmaq were gradually pushed into less productive areas with more limited access between interior and coast. The Tribe's traditional political structure and its internal authority helped it maintain social cohesion, even as free movement was drastically curtailed, but for all practical purposes, and understandably so, the loss of lands and accompanying social and economic depredations exhausted the collective. While the fur trade increased in the peninsula after the Loyalist settlement, the Mi'kmaq could not compete with poor settlers, who dominated trapping until the decline of the trade in the 1860s (Gwyn 2003).

On March 19, 1842, the government turned from the vague promises in the treaties to a permanent solution: the "Act to Provide for the Instruction and Permanent Settlement of the Indians" (Nova Scotia Legislature 1842). This act formalized what had already developed piecemeal, the establishing of reserves for Mi'kmaw people. This was in no part a recognition of tribal sovereignty, even in a limited way. Its first clause referred to "the Education and Civilization of the Aboriginal Inhabitants of this Province, and for the preservation and productive application for their use of the Lands in different parts of this Province, set aside as Indian Reservations"; and its measures were related to agricultural development, religion and education. This was the language of assimilation – consistent with the aspirations of the Jesuits and their own nation's plan for Nouvelle-France more than 200 years before.

At the time of the Act, the relatively isolated interior was a cultural refugium where Mi'kmaw could still hunt, gather and fish freely. On Fairy Lake, now a small enclosed bay on Kejimkujik Lake, a family who had traditional hunting grounds had settled and, as fur trapping was poor, had planted gardens in order to supply themselves for the winter (Chute and Speck 1999). On this evidence, the newly appointed Commissioner for Indian Affairs, Joseph Howe, convinced the government to set aside lots for farming purposes, ranging in size from 45 to 100 acres for 10 Mi'kmaw families who lived in the area: Louis Alexis (Luxey), Francis Charles, Peter Glode and his brother Francis Glode, John Jeremy and Jeremy's son-in-law, James Lewis, Stephen Labrador, Joseph Peter (Peale) and his brother Abraham Peter, and John Pictou (Chute and Speck 1999, 512). Several of these families actually did break some ground here, and Howe, for one, had seed potatoes delivered in June, 1843 (Henderson 1843).

Some of these family associations with the reserve were still present at the end of the century. In 1914 the ethnologist Frank G. Speck created a map of individual Mi'kmaw hunting territories on the peninsula from information provided by two Mi'kmaw informants (Chute and Speck 1999; Speck 1922). At Kejimkujik Lake he named Jack Glode and son Peter, and in the lands south of there, Francis Charles – both associated with families at Bear River on the Bay of Fundy – and on Medway Lake and part the upper Medway River, Stephen Bartlett (Wisa'u). Wisa'u (1819–1902), who lived near the end of his life on the Gravel Pit Reserve near Yarmouth (Whitehead 1974; 2003), was the chief of families in Shelburne and Yarmouth Counties. He used the English name Stephen Bartlett and, sometimes, included a Mi'kmaw name, *Laksi*, that was spelled various way by transcribers (e.g., Luxie, Luxy, Luxey) (Whitehead 2002, 28–29).

The lots that the government defined, as well as the boundaries of the hunting grounds, would not have represented real property to the Mi'kmaw inhabitants, but rather affirmed their common interests in the region. The southwestern part of the peninsula had river access to the interior on all sides, and families could easily shift to habitations with access to coastal or interior resources when the situation or seasons demanded.

From this evidence, it seems most likely that the people depicted in the glyphs were from one of these families, probably there while hunting, fishing or gathering, or traveling between camps. As this settlement was less than 10 miles west of the Medway River glyphs, it therefore seems most appropriate to consider them not in terms of representations of Mi'kmaq art and society of the colonial period, but rather as part of the deep surroundings of this locality, when material and social flows related to events associated with early colonization eventually resolved in the acts of one or two carvers over a period of years, got down on hands and knees and carved pictures in the slate.

The first carver was probably a hunter, well acquainted with the habits of moose, as the separate groups here depict them in seasonal behaviours. Moose are relatively solitary animals, except during the winter, when loose aggregations may form. They separate in the spring, and bull moose may sometimes be trailed by younger bulls. Males all avoid cows and their yearlings. By the autumn rut, however, the bull moose are solitary and spend their time searching for females, who will call into the forest. When the bull and cow make contact, the bull will display and drive or follow the female before mating. If there is another bull nearby, they will frequently lock horns in battle (see Geist 1963).

Being familiar with this behaviour, hunters would walk the shorelines of lakes, rivers and wetlands where the cows would be, using moose calls in the hope that it would draw a bull out of the deeper forest. It was therefore probably a hunter who waited here and carved these moose. This would also account for the outlined heart in one bull; the most effective (and respectful) way of killing is a single shot through the heart.

A second carver created the family scene and the dated hunting scene around the moose, and they seem to have been done at different times. The man with the dogs may be the carver himself, as November is in moose hunting season. The woman in the family scene, however, displays a cross and frond, which suggests the Catholic celebration of Palm Sunday, the week before Easter in the spring. While the most prominent and well-known Catholic rite for the Mi'kmaq is St. Anne's Day, July 26th, the Fairy Lake families would have also taken part in other feast days, as they attended the chapel in the nearest colonial settlement, Caledonia (see Luxy 1848).

Now that the slate rocks are repopulated with people, at least in our descriptions, we may be able to understand more clearly the nature of the second carver's choices. The deportment of the family, formal clothing and possession of a valued horse convey the new and more individualizing world of property – and, by their forthright gestures and facial expressions, a look of confidence and even optimism. This should not be surprising. The Tribe had spent more than 200 years responding to the incoming technologies and ideas as they would any incursion or opportunity, weighing the risks and benefits in the situations as they happened. In the years between 1842 and 1844, the creation of the Fairy Lake reserve might have encouraged these specific families to think they had gained some security in their hunting and gathering grounds. They were still somewhat remote from the expanding colonial settlements on the coast – and this confidence may have induced one of them to enhance the images of the moose, probably made by an ancestor, or even the same carver years before, on these rocks.

After the events

The boost that the settlement at Fairy Bay might have given its inhabitants was brief. In the world of the 1840s, just like in the world of the early 17th century, the ostensible intent of the colonists was to turn the Mi'kmaq into farmers but, in actuality, they rather wished they would simply disappear. They offered vaguely defined marginal property with sterile soils and gave them very little support as surrounding settler communities encroached on their lands and took their increasingly scarce resources with impunity.

Mi'kmaq people resisted as best they could, as the numerous letters and petitions to the government in the 18th and 19th centuries attest. In a letter to the Provincial Secretary, Joseph Howe, for example, Abraham Gesner, Commissioner of Indian Affairs, appealed for a land survey to help one beleaguered Mi'kmaq family:

> The Luxie family have long occupied a small tract on the Shelburne River at the head of Long Island, and by their labor it has been rendered capable of yeilding (sic) a rent; but the encroachment made by the neighbouring settlers threaten to deprive them of the last acre of their inheritance.
>
> (Gesner 1848)

In Queens County, Chief Louis Luxy petitioned more than once for relief for tribal members struggling through the winter, asking for food and agricultural supplies and better access to the Fairy Lake reserve, which had no road to connect it to the nearest settlement at Caledonia (Luxy 1848).

Such evidence of the well-documented struggles of the Mi'kmaq through this period may conjure up a dystopian world, but this seems at odds with an apparent sense of optimism evident in the tiny glyphs on the Medway River or, at least, the positive, resolute nature of the people. Whichever alternative an interpreter chooses, however, one thing is clear: the carvers pictured what they did in moments of *their own choosing*, without the taint of imposed values, traditions or general conditions that they could neither experience directly nor predict.

The Fairy Lake reserve persisted for several decades, but while Mi'kmaw people were adept at planting food crops (Parnaby 2008, 76–77), the land was not suitable in the first place for agriculture, as early frosts frequently killed off the potatoes and wheat. With these predictable failures, the dwindling of game from sustained hunting and trapping, and increasing incursions by settlers and loggers, families moved away. The province eventually took the land back in 1918 when they assumed it was unoccupied – apparently, some of the men were fighting in World War I and the families were in the United States at the time (Chute and Speck 1999, 539). Even then, after the Mi'kmaq were forced to give up the Fairy Lake reserve, their people at Bear River and Annapolis and Queens Counties still camped and hunted there (Chute and Speck 1999, 515).

The first dam flooded the petroglyphs soon after Creed's recording in 1888. By the first decade of the 1800s, most of the good logs had already been taken near the coast and the loggers had begun to move further into the interior along the Medway and other major rivers, building the driving dams that would hold back spring waters until they were needed to float logs downstream (Johnson 1986, 81). Soon after the 1983 recording, the dam on the Medway River was repaired and the glyphs disappeared again underwater, probably for many more generations.

Families who traditionally hunted and fished up the Medway River and at Kejimkujik probably have stories of their ancestors there, and their own personal traditions regarding the petroglyphs. While there is no essential reason for this community or any other to care about the characterization of rock art here or its background, the analysis at least treats ancestral life at a scale relevant to the people who lived it and to their descendants, who continue to assert and defend the rights and interests they inherited. It therefore offers no guidance or solution – only the possibilities that the situation describes. If it has value, it is because it uses source materials at the scale of the problem and attempts to stay close to the site and its surroundings. This is a task that history and historical archaeology take on, using intensely detailed historical and genealogical research, in what ethnologist Ruth Whitehead has done in Nova Scotia, to retrieve people "lost in time" (Whitehead 2002, 43) – and this emphasis on actual people, localities and events should be an essential part of any archaeology, including rock art, despite the absence of supporting information at so many sites.

Beyond the immediate family considerations, the broader interpretation of these and other petroglyphs among Mi'kmaq people depends on who sees them, and which ideologies they choose to value among the range of interests and alliances available to them (see Poliandri 2011).

The original Jesuit project may have contributed to what the carver created in the 1840s. From its beginnings, Catholic education emphasized the teaching of indigenous people through pictures, and learning through pictures eventually transformed them into witnesses of their own lives. This individuation was by no means restricted to the Mi'kmaq or to missionaries. From its beginnings in the 15th century in Italy, pictures began to take part in the capitalism of perception, by which elites use the natural and social power of salience to reinforce ideology and ensure social control and, as the 19th century progressed, encourage material desire among ordinary people.

Ironically, such individualism among indigenous people in 19th century Nova Scotia and elsewhere created the opposite condition, one apparent today. Continuing social depredations, including the creation of reserves and the establishment of assimilation schools, eliminated local social diversity, which was centered around extended families who shared homelands across the peninsula, and they coalesced around the common traditions they could still maintain. Perhaps more attention in general to life in these small communities, preserved in the lands they occupied, might provide a helpful correction to more general assumptions about this past.

Images as situations

In concluding our discussion of rock art, our focus on the act of an artist changes the field of possibilities for interpretation in rock art. Before the image is the situation and the act, and so the image is an analogue of a situation's resolution, an outcome. At least one of the carvers on the Medway River was hunting for moose and scratched out some pictures on this smooth slate rock along the banks while pausing there. While we seize on an image and wonder what it means, it is important to consider that the image is some realization of the maker's own past, before and during production.

This is why we turn from the search for meaning to look at the artist's surroundings. If the reality of an image is in the making of it, and creation takes place in a situation with physical and ideological aspects, there must be more to the process than retrieving images from one world for matters in another.

Imagining rock art as a situation, consider the act more closely, as a direct encounter. In every cave or shelter and on every rock wall, there was a *first artist*. This painter or carver chose sites out of the surroundings and composed and placed their images at will. Even if the choice was predetermined – for example, because the place was an abode of spirits – this artist had the freedom to work within the topography of the surface. This was not so for the followers. Each successive artist worked with an awareness of the images already there. This sensory and ideological process would go on, in a cascade of individual acts and situations, until circumstances changed and there was no longer any need, opportunity or desire to make pictures there at all.

When one looks at the art this way, different questions emerge. The first is about the decision to make pictures in the first place. Art is an irruption in the habitude of ordinary life, and its display changes the distribution of salience in the perceived world. This means that image-making is never neutral after the first act. Like the naming of a thing, the first painting or carving marks a rock with silent but visible intent and shapes what follows – in environment and ideology.

In the end, the visible results of all the work on a rock are arranged for all to see, and they show how each situation resolved – especially a particular artist's regard for the visual integrity of the image. Where overlays occur, one can speculate about whether the individual artists were aware of or cared much about images already in place, and wonder if the act was intentional defacement, an elaboration or transformation of ideas or sheer indifference. One thing is clear, however: with every act, there was increasing suggestiveness and decreasing space.

Introducing complexities at the site in a study of rock art when the norm is a quick leap from picture to the analysis of meaning and function may be difficult, but it is important to consider picture interpretation in the environment of the act. This ecological approach to art may have some benefits for research. If we are resolute in staying close to the ground, what is remarkable in rock art is how similar the situations were that all artists faced in matters of display. Over at least 45,000 years, and no doubt before that, humans have created visual things and responded to them directly and simply, without the need for guidance. While all the image-makers in the past lived in different societies and circumstances, they shared the bodies and senses of their common ancestors in a singular world. And because this physical world moves at a vastly slower rate than human life, we often find that some of the surroundings and circumstances in situations of production still exist.

Thinking about situations opens the search for clues to an artist's intent in ways impossible through the application of meanings. We can be certain, for example, that the makers of handprints in the Sulawesi caves and the painter of the rhino in Lascaux actively intervened in the small world of the art that already existed at these sites, and that thousands of years later, in 1844, a

Mi'kmaq carver would do the same on the Medway River. These interventions are not obvious in an isolated picture, which is so easily a product and a container of meaning, and crucially, remote analysis and interpretation overlook the possibility that these artists wanted to say something specific to their own experience through their own creations.

This makes sense across the whole of art, because from the first daub or spray of pigment, or the turning of a tool, the artist is a director of attention. This has to do with the simplest aspect of visual perception. A picture is silent, and so if communication is a goal, the maker may attempt to lead a spectator's eye through the depiction to get to the source of its meaning and intent. As the act of composition combines perception, action and ideology, a situational approach should therefore apply across the conventional boundaries of art – time, culture, society and medium – because its focus is not on meaning, but on the shape that meaning takes.

We now move from outside environments across another abyss, the disciplinary space between archaeological or anthropological studies of art and art history, and look at European paintings. In comparison to rock art, information about the arts of Europe is overwhelming, with more than 2000 years of scholarship. For an archaeologist in this strange land, it may seem impossible to expose the spare dynamics of situations. Yet, these painters also stood in front of their walls and canvases, and revealed their intentions in their acts and so thinking about the pictures as *their* interactions may provide firmer ground for the interpretation of the results.

References

Akins, Thomas B., ed. 1869. *Selections from the Public Documents of the Province of Nova Scotia*. Halifax: Charles Annand.

Bachelard, Gaston. 1994. *The Poetics of Space*, Translated by Maria Jolas. Boston: Beacon Press.

Bakker, Peter. 1989. "The Language of the Coast Tribes is Half Basque": A Basque-American Indian Pidgin in Use between Europeans and Native Americans in North America, ca. 1540–ca. 1640. *Anthropological Linguistics* 31:117–147.

Bemiss, Samuel M. 1957. *The Three Charters Of The Virginia Company Of London. With Seven Related Documents; 1606–1621*. Jamestown 350th Anniversary Historical Booklet Number 4. Williamsburg: Virginia 350th Anniversary Celebration Corporation.

Bordin, Guy. 2015. *Beyond Darkness and Sleep: The Inuit Night in North Baffin Island. Arctique* 10. Société d'études linguistiques et anthropologiques de France. Paris: Peeters.

Brantingham, P.L., and P.J. Brantingham. 1995. Criminality of place: Crime generators and crime attractors. *European Journal of Criminal Policy and Research* 3:5–26.

Buckley, Theodore Alois. 1851. *The Canons and Decrees of the Council Of Trent. With a Supplement, Containing the Condemnations of the Early Reformers, and Other Matters Relating to the Council. Literally Translated into English*. London: George Routledge and Co.

Burley, David. 1981. Proto-Historic Ecological Effects of the Fur Trade on Micmac Culture in Northeastern New Brunswick. *Ethnohistory* 28:203–216.

Chute, Janet E., and Frank G. Speck. 1999. Frank G. Speck's Contributions to the Understanding of Mi'kmaq Land Use, Leadership, and Land Management. *Ethnohistory*:481–540.

Dekoninck, Ralph. 2017. The Founding of a Jesuit Imagery: Between Theory and Practice, between Rome and Antwerp. In *The Acquaviva Projects: Claudio Acquaviva's Generalate (1581–1615) and the Emergence of Modern Catholicism*, edited by Pierre-Antoine Fabre and Flavio Rurale, 331–346. Boston: Massachusetts Institute of Jesuit Sources, Boston College.

Eck, John E. 1997. Preventing Crime at Places. In *Preventing Crime: What Works, What Doesn't, What's Promising.* A Report to the United States Congress. Prepared by Lawrence W. Sherman, Denise Gottfredson, Doris MacKenzie, John Eck, Peter Reuter, and Shawn Bushway. Prepared for the National Institute of Justice in collaboration with members of the Graduate Program, Department of Criminology and Criminal Justice. Baltimore: University of Maryland. https://www.ncjrs.gov/works/.

Ewers, John C. 1955. *The Horse in Blackfoot Indian Culture, with Comparative Material from Other Western Tribes.* Smithsonian Institution, *Bureau of American Ethnology Bulletin* 159. Washington: Government Printing Office.

Fleming, Alison C. 2019. Jesuit Visual Culture: Communication, Globalization, and Relationships. *Journal of Jesuit Studies* 6(2):187–195.

Geist, Valerius. 1963. On the Behaviour of the North American Moose (Alces Alces Andersoni Peterson 1950) in British Columbia. *Behaviour* 20:377–415.

Gesner, Abraham. 1848. *Letter from Abraham Gesner*, Commissioner of Indian Affairs, to Joseph Howe, Provincial Secretary, May 5, 1848. Commissioner of Public Records — Indians series Nova Scotia Archives RG 1 Vol. 431 No. 47.

Gwyn, J. 2003. The Miíkmaq, Poor Settlers, and the Nova Scotia Fur Trade, 1783–1853. *Journal of the Canadian Historical Association* 14:65–91.

Henderson, Andrew. 1843. *Letter from Andrew Henderson to Joseph Howe*, Provincial Secretary, June 21, 1843. Commissioner of Public Records – Indians series Nova Scotia Archives RG 1, Vol. 432, 221–222.

Ingstad, Helge, and Anne Stine Ingstad. 2001. *The Viking Discovery of America: the Excavation of a Norse settlement in L'Anse aux Meadows, Newfoundland.* New York: Checkmark Books.

Johnson, Ralph S. 1986. *Forests of Nova Scotia. A History.* Halifax: Nova Scotia Department of Lands and Forests.

Keyser, James D. 1979. The Plains Indian War Complex and the Rock Art of Writing-on-Stone, Alberta, Canada. *Journal of Field Archaeology* 6:41–48.

LeHuenen, Joseph. 1984. The Role of the Basque, Breton and Norman Cod Fishermen in the Discovery of North America from the XVIth to the End of the XVIIIth Century. *Arctic* 37:520–527.

Lescarbot, Marc. [1606] 1928. *Nova Francia, a Description of Acadia*, translated by P. Erondelle, 1609. New York: Harper & Brothers.

Luxy, Louis. 1848. *Letter to the Honourable Members of the House of Assembly*, February 14, 1848. Commissioner of Indian Affairs Nova Scotia Archives MG 15 Vol. 4 No. 60.

Mallery, Garrick. 1893. *Picture-Writing of the American Indians.* Tenth Annual Report of the Bureau of Ethnology to the Secretary of the Smithsonian Institution, 1888–1889. Washington: Government Printing Office.

McCall, John E. 1948. Early Jesuit Art in the Far East IV: In China and Macao before 1635. *Artibus Asiae* 11:45–69.

Mead, Margaret. 1937. *Cooperation and Competition among Primitive Peoples.* New York: McGraw-Hill Book Company.

Medina-Alcaide, M.Á., D. Garate, I. Intxaurbe, J.L. Sanchidrián, O. Rivero, C. Ferrier, Dolores Mesa, Jaime Pereña and Iñaki Líbano. 2021. *The Conquest of the Dark Spaces: An Experimental Approach to Lighting Systems in Paleolithic Caves.* PLOS ONE 16(6): e0250497. https://doi.org/10.1371/journal.pone.0250497.

Molyneaux, Brian. 1983. *The Study of Prehistoric Sacred Places: Evidence from Lower Manitou Lake.* Archaeology Paper 2. Toronto: Royal Ontario Museum.

Molyneaux, Brian L. 1989. Concepts of Humans and Animals in Post-contact Micmac Rock Art. In *Animals into Art*, edited by Howard Morphy, 193–213. London: Unwin Hyman.

Mulan, Father Elder, trans. 1914. *The Spiritual Exercises of St. Ignatius.* New York: P.J. Kenedy.

Murray, Jean E. 1938. The Early Fur Trade in New France and New Netherland. *Canadian Historical Review* 19:365–377.

Nova Scotia Legislature. 1842. *Act to Provide for the Instruction and Permanent Settlement of the Indians.* Commissioner of Public Records – Indians series Nova Scotia Archives RG 1 Vol. 432, 9.

O'Connell, Heather. 2017. What Happens in the Shadows: Streetlights and How They Relate to Crime. Houston: Rice University Kinder Institute for Urban Research. https://doi.org/10.25611/41qx-u9sm.

Parnaby, Andrew. 2008. The Cultural Economy of Survival: The Mi'kmaq of Cape Breton in the Mid-19th Century. *Labour* 61:69–98.

Perkins, Chloe, Rebecca Steinbach, Lisa Tompson, Shane Johnson, Ben Armstrong, Judith Green, Chris Grundy, Paul Wilkinson and Phil Edwards. 2015. *The Effect of Reduced Street Lighting on Road Casualties and Crime in England and Wales: Controlled Interrupted Time Series Analysis.* (Public Health Research, No. 3.11.) Southampton (UK): NIHR Journals Library.

Poliandri, Simone. 2011. *First Nations, Identity, and Reserve Life. The Mi'kmaq of Nova Scotia.* Lincoln: University of Nebraska Press.

Sartre, Jean-Paul. 1948. *The Psychology of Imagination.* Secaucus: The Citadel Press.

Simpson, J., and E. Weiner, eds. 2009. *Oxford English Dictionary.* Oxford: Clarendon Press.

Speck, Frank G. 1922. *Beothuk and Micmac.* New York: Museum of the American Indian, Heye Foundation.

Thwaites, R.G., ed. 1896a. *Travels and Explorations of the Jesuit Missionaries in New France 1610–1791.* Vol. 1. Cleveland: The Burrows Brothers Company.

Thwaites, R.G., ed. 1896b. *Travels and Explorations of the Jesuit Missionaries in New France 1610–1791.* Vol. 49. Cleveland: The Burrows Brothers Company.

Tilley, Christopher. 1994. *A Phenomenology of Landscape: Places, Paths and Monuments.* Oxford: Berg.

Tilley, Christopher, and Kate Cameron-Daum. 2017. *Anthropology of Landscape: the Extraordinary in the Ordinary.* London: UCL Press.

Triandis, Harry C., Robert Bontempo and Marcelo J. Villareal. 1988. Individualism and Collectivism: Cross-Cultural Perspectives on Self-Ingroup Relationships. *Journal of Personality and Social Psychology* 54:323–338.

Whitehead, Ruth. 1974. *The Micmac Ethnology Collection of the Nova Scotia Museum. Curatorial Report Number* 25. Halifax: Nova Scotia Museum.

Whitehead, Ruth Holmes. 2002. *Tracking Doctor Lonecloud.* Fredericton: Goose Lane Editions.

Yellen, John. 1990. The Transformation of the Kalahari !Kung. *Scientific American* (April):96–105.

7 Legend of the true cross

Pictures in churches

In an ecology of art, a picture is both a shelter for ideology and an environment within a larger world of action that links directly to its surroundings, combining physical senses and social experience. Without a sense of a brute nature in this environmental interaction, however, ideologies have no grounds, and can only interrogate and celebrate themselves. If bringing in the physical world seems problematic for those who invest in meaning, perhaps the solution is not to leave empirical matters to science, but to put culture under scrutiny rather than nature, and consider culture as a nature that sees itself. With this turn, we can ground sensibility and focus on the actions of artists in the most spiritual of places.

Christian cathedrals and churches often strike the senses, striving above the horizon with imposing towers and spires that announce their presence, or thick and sturdy blocks with great domes, set firmly in the earth. These stony enclosures provide the shelter and sanctuary imagined for the caves painted during the Paleolithic in western Europe.

Church interiors are intended to keep out the world. Entry is through a small side or rear door into an open space, lit naturally with a few grills or leaded glass windows; in this open nave the congregation gathers in pews, facing an altar.

As if to compensate for sensory deprivation, ceremonies may deeply involve the senses, with colourful stained glass windows, candlelight, the perfume of censers, and singing and music. It makes the performative speech and rituals within seem extraordinary, even though they are from the same sources as ordinary life.

The sensory integration of culture and nature is not the same, however, when the subject is the art. By the time of the writing of the Bible, religious scholars were unsure about picturing beliefs. The Second Book of Moses, which relates the story of the great exodus of Hebrew people from Egypt, makes a proscription against imagery quite clear: "Thou shalt not make unto thee any graven image, or any likeness of any thing that is in heaven above, or that is in the earth beneath, or that is in the water under the earth" (Exodus 20:4 KJV). Several chapters later, however, God bids Moses to build an ark

DOI: 10.4324/9781351053105-7

and set on it a mercy seat, with "two cherubim of gold" on the ends (Exodus 25:18). And when Solomon, the son of King David, eventually builds the temple, it is said that he places two cherubim made from olive wood in the inner sanctuary and has cherubim, palm trees and open flowers carved on the walls of inner and outer rooms (1 Kings 6: 23–36).

The lack of clarity about the rightness of religious images began centuries of ideological contortions that unsurprisingly extended into Christianity, as religious leaders and philosophers tried to deal with the Bible's strictures while continuing to reinforce its message through this most effective of media. In the second of two letters that Pope Gregory I (ca. 540–604) wrote in 600 CE to Serenus, the bishop of Marseille, after learning that iconoclasts had destroyed images in churches, he admonished the Bishop, writing: "Idcirco enim pictura in ecclestis adhibetur, ut hi qui litteras nesciunt saltem in parietibus uidendo legant, quae legere in codicibus non ualent" ("Pictures are used in churches so that those who are ignorant of letters may at least read by seeing on the walls what they cannot read in books") (Duggan 2005, 63).

This epistle, which clearly addressed a spiritual gulf between the intellectual elite and the masses, seems pragmatic, and it was visited with caution. It was the diversity of scholarly readings of the scriptures, and the fine line between worshipping through a picture and adoring its image, that caused frequent bouts of iconoclasm, especially in the Byzantium Church, which was also wary of the rise of Islam. During the reigns of Gregory's successors, Gregory II and Gregory III, Emperor Leo III of Constantinople vacillated between defending their use, as in his epistle to the Umayyad Caliph 'Umar II (Jeffrey 1944, 322), and placing edicts against their veneration (see, e.g., Turner 1990).

John of Damascus (ca. 675 – ca. 749), defender against iconoclasm, provided a Christian rationale in 730 that would suffice for virtually any symbolic understanding of pictures over succeeding centuries. Images had immanent spiritual power, because God, "not wishing that we should be in ignorance of spirits, clothed them in symbol and figure, and in images like our nature, material forms visible to the mind in mental vision", and he then asks: "These we put into form and shape, for how were the cherubim represented and described in image?" (Allies 1898, 100).

As the centuries passed, the controversy about pictures remained, but vision changed from an uncomplicated process that matched pictures, mental visions and unformed spirits, to one that focused more on seeing through the senses. This decidedly modern view is clear when St. Bonaventure (1221–1274) reaffirms the church's orthodoxy on pictures in three statements:

1 They [images] were made for the simplicity of the ignorant, so that the uneducated who are unable to read Scripture can, through statues and paintings of this kind, read about the sacraments of our faith in, as it were, more open scriptures.
2 They were introduced because of the sluggishness of the affections, so that men who are not aroused to devotion when they hear with the ear

about those things which Christ has done for us will at least be inspired when they see the same things in figures present, as it were, to their bodily eyes. For our emotion is aroused more by what is seen than by what is heard.

3 They were introduced on account of the transitory nature of memory, because those things which are only heard fall into oblivion more easily than those things which are seen.

(Duggan 2005, 73)

It seemed that pictures did not simply invite viewers into the world of the spirit. Despite the anarchic sense about them, they had specific qualities of immediacy, salience and persistence that had some distinct advantages in the teaching of the Bible. Church fathers therefore remained convinced that the reason why it was so easy to convey the sensory richness of Bible stories in depictions and so difficult to teach their moral lessons was a matter of education. Well into the 16th century, when Ignatius of Loyola (1491–1556) wrote his Spiritual Exercises, he rationalized this ancient conviction in the contemplation of the Nativity, using the "five senses of the imagination" (Mulan 1914, 65) to bring this formative event into reality. In this exercise, the first step was "to see the persons [Mary, Joseph and the maid and, then, the child Jesus] in the sight of the imagination" (Mulan 1914, 62). A priest might therefore expect that the masses would achieve this state after learning the scriptures through the spoken word. And this led to centuries of missionary work by the Society of Jesus.

The enduring conviction that pictures stimulated the imagination through applied meaning prepared the groundwork for the modern history of art and its traditional home in representations. It was quite ironic, however, that these defenses of religious art outlined precisely why pictures are not particularly effective as teaching tools. Their primary appeal is to the senses, emotions and imagination, and so they resist social control, just like the naturally salient features of their surroundings. Looking back at the Bible's indecisive statements about iconoclasm, it is clear that there has always been a visceral church, one that thrives on the emotions of belief through the stimulation of the senses or their denial, and that it has an equally long history. The ideological church may overwhelm these base behaviours with its teachings, but it cannot extinguish the natural energies in its shelters and presentations.

The enduring result over centuries of scholarship has been a push and pull between intellectualism and the sensory displays that impress the masses. From this, we may conclude that the significance of pictures is not only a philosophical problem but also an environmental one. Sensory effects do not compete with the religious, social or political aspects of display and communication; they are the energies and forces involved in them.

Thus, our subject. If we presume that the situations of Paleolithic painters and the makers of rock art are commensurate, this must also be so for the painter in the sensory terrain of a church. We therefore move from the stony ground of the outside world to the basilica of San Francesco, a late Medieval church in Arezzo, Tuscany, and into its chancel, where the Tuscan artist, Piero Della Francesca (ca. 1415–1492), painted frescoes on the walls. The physical constraints of the church environment provide the stable grounds from which to search for evidence of artists' intents in the work and what this suggests about the situation at the time.

Piero's anomalous legend

The basilica of San Francesco was begun near the end of the 13th century by Franciscans in Arezzo, a small city in Tuscany. Sometime in the 1440s, the Bacci family, wealthy merchants in Arezzo, commissioned the decoration of the chancel walls with illustrations of a well-known legend about the "true cross" on which the Christ was crucified.

The story of the cross is best known from a popular hagiography, the Legenda Aurea (Golden Legend), as compiled by Jacopo da Varagine (ca. 1230–1298), archbishop of Genoa, in the 1260s. It follows the journey of the holy wood from its genesis at the death of Adam through an exciting history that includes the visit of the Queen of Sheba to King Solomon in Jerusalem, the burial and recovery of the cross, wars among Byzantine emperors, wars with the Persians and, eventually, the return of the holy wood to Jerusalem (see Baert 2004). It was immensely popular at the time; Robert Seybolt (1946) recorded 173 editions of Varagine's book during the 15th century. In Piero's time, the readers were limited to clergy and literate members of the laity until Niccolo Jensen published the first Italian edition in 1475. With its mixture of Biblical accounts, apocrypha, legend and history, however, the story was probably shared more widely as an oral tradition.

The Bacci commission went to Bicci di Lorenzo (1373–1452), a Florentine painter, but when he died in 1452, the chancel walls were still bare. To complete the work, the Baccis turned to Piero Della Francesca, an Aretine painter from the village of Borgo Santo Sepolcro, and he seems to have completed the task by 1466.

In the rather small chancel space below the vaults depicting the heavens, he arranged episodes from the Legend in three horizontal registers that run across the side walls and the space around the window on the back wall. The top registers of the side walls are lunettes. In these spaces, high above the floor of the church, Piero begins and ends the story: the failing and death of Adam on the right, and the climax of the story, when the Emperor Heraclius brings the wayward cross back to Jerusalem on the left. In between, however, Piero's organization is obscure, as his visible sequence of episodes does not follow the story as a progression from one wall to the other, but rather, jumps back and forth. This is not at all like Varagine's story, a running commentary read page-by-page and, so, not at all as the ancient church fathers intended, that the picture-story could be read like a book. Indeed, scholars who see a logical sequence need to guide viewers through the story, often numbering the individual frescoes to indicate where to turn and look next.

It might seem odd to say, but we can only assume from the vague evidence that his contract, and his ostensible task, was to present a visual narrative of the Legend.

He clearly had to meet the expectations of the friars and the Bacci family donors, and he had two great examples to follow. Giotto di Bondone (1267–1337) painted his stories of the Life of Christ and the Life of the Virgin more than 150 years earlier in the Scrovegni Chapel in Padua, and Agnolo Gaddi (ca. 1350–1396) painted his own version of the True Cross in the chancel of Santa Croce, Florence, near the end of that century. And even the most plodding artist could follow Varagine's brief and summary description of the events. Yet, there is no record that Piero's treatment caused any stir in Arezzo, which suggests that his audiences were at the very least content. Indeed, Vasari (1511–1574), who was born in Arezzo, went on to paint for the Medici family in Florence and wrote the first book of art, artists and their biographies, *The Lives of the Most Excellent Painters, Sculptors, and Architects*, was pleased with the result ([1543] 1912, 20).

> Here there are many beautiful conceptions and attitudes worthy to be extolled; such as, for example, the garments of the women of the Queen of Sheba, executed in a sweet and novel manner; many most lifelike portraits from nature of ancient persons; ... and a peasant who, leaning with his hands on his spade, stands listening to the words of S. Helena while the three Crosses are being disinterred with so great attention, that it would not be possible to improve it.

This rather bland acceptance will not do, however, for those who want to find a narrative logic on these walls. The attempt makes sense: Piero clearly induced flows within and between episodes and began and ended appropriately, suggesting that the whole must be a visual journey from the top of one side to the other. It is not only a question of finding the best route, however, as there is an additional problem. Across from a depiction of one of the most significant events, when the Emperor Constantine has a dream that convinces him to carry the banner of Christianity into battle, Piero inserts a scene of the Annunciation and, so, takes the viewer outside the story. Thus, with no other evidence to go on, it seems that Piero, in work over a decade, somehow fails, does not bother or declines to deliver either conventional narrative or logic to this popular story.

This puzzle, and the framing of the problem as a question of how one looks at the frescoes from the floor of the chancel, suggests that the Legend is a good place for a situational analysis.

It might seem obvious that we should just press on and act as one spectator among others in the church terrain, following a routine established in the 15th century, and attempt to see the walls as the artist did. Yet, as an interpretive approach, this immediately brings up a variety of issues related to conceptions of knowledge and the constraints of individual and social relativism. The choice to press on usually means working within the shelter of existing narrative traditions, institutionalized over centuries in art history and, more recently, anthropology. Such institutions are able to gather huge amounts of information and marshal many scholars and students to the task of understanding history and social processes, and bring this experience to bear, in this case, on the frescoes. From this view, one can imagine the church of San Francesco as a landscape, within which clergy, congregations and artists all gather and move within valorized social processes.

While the church and the surrounding community may be gathered in such a representation, it is more than six centuries since the day Piero took on the commission to decorate the chancel. It seems impossible to imagine him severed from this future, when he was not an early master of an art form and intellectual movement in a city-state and society that was not a part of Italy. It is speaking a language that has yet to be developed.

The alternative is to work at the scale of individuals rather than the scale of ideas, as this is the shared experience of all interpreters, however great their imaginations and the models, representations and workers they can marshal. The question is whether it is possible to learn something about the situation in the mid-15th century in Arezzo by occupying and acting in the same physical grounds as the artist, clergy, patrons and congregations. Outside the shelter of the institution and its authority, what exists are the material traces left from the painters' acts – the analogue imagery – and the physical terrain of the church, its architecture and furniture. What we propose is that from using these features in this locality, we may be able to find indications of the artist's personal intent and, with this grounding, move out into the surrounding environment to account for these actions using the traditional evidence of life at the time.

The situation from the ground

Standing in the basilica today and looking up at Piero's extraordinary displays is like standing at the edge of a cultural abyss. What he painted is there on the wall. How the images appeared to him or his assistants, or to the clergy, patrons and congregations, or, for that matter, to contemporary art historians and archaeologists, is another problem. Issues of appearance and reality in the subdued light of the basilica are no different than in the darkness of a Pleistocene cave.

This problem is especially so in a sacred place, where the lure of appearances seems natural. A church is a building constructed to unite and direct the senses and imagination to a spiritual world, and the religious scenes on the walls act as gestalts, promising a wholeness much greater than the people and events they depict. Indeed, after almost 600 years of attention, the paintings in the chancel in San Francesco exist with all the other artworks of the period in a representational afterlife, a bewildering complex of ideas from philosophy, history, psychology, sociology and anthropology and, most recently, biology, each field with its own way of looking at what artists laid down in pigment on the walls.

If a picture has only the past of a present construction, then what the artist creates is deprived, an unintelligible mass in a liminal state waiting description. Sartre saw this essential poverty in images: "No matter how long I may look at an image, I shall never find anything in it but what I put there" (1948, 19). Looking back to writings before the Renaissance, however, there is a clarity about the issue of pictures that suggests a different approach than endless first-person descriptions of experience. Whether these scholars defended images or were iconoclasts, they all understood the earthly impact of pictures – their directness, salience and persistence – and knew that individual people saw in them what they wanted. The church fathers believed that it was their task to find ways to educate the attention of the masses to representations suitable to their beliefs.

During the 15th century, the efforts to manage attention to the meanings in pictures became the direct concern of artists, who worked with patrons under commission, still officially under the aegis of the Church. Their practical efforts to direct perception are clear in the studies of the geometry of optics and visual perspective.

These ancient concerns relate to what is common in pictures, despite the vicissitudes of belief and imagination – the act of looking at them. Over thousands of years of human history, and despite the constraints of phenomenology, people have readily recognized pictures and other forms of depiction alien to their own experience without the need of instruction. This mildly anarchical implication, that perception exists outside the capacity and control of governing representations, fits with the pragmatic views of philosophers in the 20th century, such as Dewey and Mead, who held that social life exists in the act, the moment, the dynamic interaction between individual and environment. While philosophers have always recognized the freedom of perception, the freedom from representation that these pragmatic views afford opens up the study of perception to a wider field of possibilities in the study of art. Here, it is the idea of art as display, the visual salience and sensory richness that are the active forces in the drawing power of visual things.

The matter of display

Unlike a concept of art, display has no historical moment. Its reflective surfaces provide the energy of delusion (Shklovsky 2007) for image-makers, as it has since hominins began to choose what and how to use coloured pigments to attract the attention of others. Fine art is the legacy of this display, rationalized and refined over centuries as a social expression.

When one considers art through display, as an ecology, there is no need to push its origins deeper into the past, along with the evolution of a symbolic mind, with every discovery of an even more ancient work. As it is, the recent depiction of a Celebes warty hog in Sulawesi, dated well before the evidence in European caves (Brumm et al. 2021), and the seemingly Neanderthal handprint in an Iberian cave (Hoffman, Standish et al. 2018) should cause a

Copernican revolution among prehistoric art specialists (cf. Moro-Abadía and Maidagan 2010). These finds are evidence of the increasingly intricate ways that hominins used display.

Having said that, working through display is not a strategy to dissolve the boundaries of art as an institution. It only acknowledges that commensurate relations exist among human beings, despite the vicissitudes of culture and its theories. Do some people believe the gods or spirits created their art or that a painting of a horse means nothing about a horse? Of course. But it is the display that first captures attention, and so offers more to find in a picture that what it means.

A crucial distinction of display from art is the nature of the interaction between perceiver and depiction. From anthropology, Gell (1998, 7) states a conventional view:

> … the nature of the art object is a function of the social-relational matrix in which it is embedded. It has no "intrinsic" nature, independent of the relational context.

In this anthropological sense, the active viewer is an agent, a proxy in a social world of models and representations that provides the content of the description and accounts for how and what this individual sees. The problem with this view is its origins in a representationalist tradition that denies the possibility of being outside its conceived world. Imagine our metaphorical dog in the company of a landscape painter, the two of them perched on the bank of a pleasant stream sketching the view, when the bank gives way and they tumble down the slope. Only the dog is allowed the brute nature of the fall. And when they shake themselves off and climb back up, only the dog carries on regardless. The artist is forever trapped in the social implications of the act.

Gell's reasoning for adhering to a social matrix was actually quite practical, as he offers a theory that aligns with other anthropological theories and so he rightly focuses on the institution and its teachings. He was actually more of a dissident, however, confident to show the conceptual weakness of the idea of culture – "Culture has no existence independently of its manifestations in social interactions" (1998, 4).

If his bond to a social matrix is overlooked, his work has a grounding in action that contributes greatly to a study of display and its situational approach. In common with Dewey and other pragmatists, for example, he rejects the characterization of picturing as a kind of symbolic language, first, because it attempts to cram one concept into the world of another and, second, because, like all symbolic approaches, it abandons the physical picture as an object of study. Indeed, like other scholars of the period, he adopts ideas from Gibson's perceptual ecology, as suggested in his critique of iconography, when he states that "A picture of an existing thing resembles that thing in enough respects to be recognized as a depiction or model of it" (1998, 25).

So while Gell is fully preoccupied with the practical mediation of art objects in social processes (Gell 1998, 6) his decentralized concept of social interaction has its grounds on the surface of the world, much like the grounds of situations, and this supports crucial elements of display.

The surface dynamics in display are thus what one is drawn to see on the spot, as opposed to what meanings might follow. If the spectre of representation is set aside, this grounding opens up the study of pictures to interactions that go beyond the traditional confines of art.

Most importantly for any study of art, the appeal of display is that it is local, keeping analysis close to the richness of the raw experience of the object or image as the maker produced it and spectators of the time took it in, and it is silent – and, thus, environmental. We will never know what Piero and his assistants had in their heads, but we do know where they walked and stood, how the light carried through the space, as they faced the situations that arose every day in front of the walls, and what they painted, which becomes a part of nature as soon as they lay the tools down.

Words fail: art without meaning

In the study of display, the silence of art is not an invitation to supply meaning, as if the artist, impoverished without language, somehow failed to deliver the response required.

As life is not all culture, art is not all meaning. The artist at work is absorbed in the task and faces a cascade of situations that are a conjunction of unawareness, tacit knowledge and strategic action.

We are mindful here of the inspirations of the Tao and the Lakota sense of wakan. Nowhere in meaningful approaches does the artist respond to affordances and other sensory provocations like other creatures, in ways that might travel from hand to surface without social mediation. And nowhere is the viewer allowed the same privilege, of acting directly without the intervention of higher concepts. Without the unacknowledged and the unspoken, however, there may be culture in an act but no humanity. When every movement and gesture is social, it is as if culture replaces God as the Alpha and the Omega (Revelation 22:13 KJV) – that is, as the totality of what can be uttered.

The idea of action without intrusive meanings is a problem for interpreters bent on narrative, but not for artists. Words often fail artists when they are forced into the retrospection. As Jasper Johns observed (Welish 1996):

> … things become "other" when you look at them backwards. Dissecting a painting is different from constructing a painting. When you look at a painting, you can't easily go back to the point at which it was begun.

Even where artists offer their own interpretations, they risk participating in an alien world outside their creations. Of course, this implies that an artist may not understand what they have created – but this is a problem of narrative, not art.

Consider the work of sculptor Rachel Whiteread, who casts the empty spaces within and around ordinary things and creates provocative objects that defy ordinary expectations of environment. In 1990, she cast the interior of a Victorian parlor in a vacant home in north London. The unsettling, enigmatic result, which she named Ghost, sparked a host of interpretations that view the piece within the context of domesticity, as in Carley's analysis (2008, 25):

> Ghost provides evidence of a domestic afterlife. The spirit of a dead interior is made manifest, as the artist reconfigures and reinvigorates the significance of a demoted, quotidian space, rendering our relationships to our habitual haunts deeply ambivalent.

Whiteread's own account, which she recalled in a number of interviews, is substantially different from this typical narrative. Of course, her language was conversational, but importantly, she finds the meaning of the work in her direct interaction with it (2014).

> … going in one morning on my own and opening the door and the sun was coming in and there was this sort of incredible light. And I saw the door and the light switch and I had this sort of eureka moment. I thought, I'm the wall, that's what I've done. I'm the wall and the viewer is the wall …

Reticence in artists about interpretation reinforces the idea that creation is not simply the re-presentation of ideas. Following the lead of artists, as opposed to interpreters, duly opens up a field of action possibilities in the study of their works and the situations they faced in production.

Outcomes and analogues

Consider two different ways of looking at a picture. On the one hand, the work is a product, the object left when the maker lays down the tools. Free of the artist, it faces into the future, open to endless interpretation, and its appearance may be copied and taken away for study as a discrete object. On the other, it is an outcome of a series of acts, an analogue environment that bears the evidence of its creation. Bound to the artist, this object is its own past, a record of the act as it happened, like the vibrations of an earthquake traced on a seismograph.

This is a crucial distinction, as each way of seeing the object offers different possibilities for interpretation. For a product, the act of creation is usually irrelevant. What counts is what the finished object represents about the society in which the artist worked. As historiography shows, such interpretation is relative to wider social processes active at the time. In one era, Vincent Van Gogh (1853–1890) is a troubled painter; in another, he is a mirror of his society.

The study of the analogue image looks the other way, towards the acts rather than the results. This approach imagines the artist at work in a physical environment, at the face of a cave or rock outcrop, wall or canvas, who shapes intentions and meanings amidst the material and social forces active at the time. This Van Gogh may be troubled, or the mirror of a troubled society, but the particular concern is how, in a single painting, he resolved the cascade of situations he experienced, seeing, thinking and acting during the task of creation, and not a single day after he laid down his tools.

In this view, a picture or sculpture is a kind of terrain, the physical ground within which the artist acted in the circumstantial flows of material and social energy when and where the work took place. During the task, the artist was a maker, using senses and kinetics mediated by experience, and an observer, who planned, watched and guided the acts. This cascade of motions and decisions produced the visible result.

As a study of the act, this approach has an environmental aspect that allows for the commensurability of two essential viewers: the artist, who shapes the image and distributes its salient features across the ground, and the analyst, who attempts to follow these salient traces in the outcome. Their spectations are generally comparable because they share physical senses, working space and affordances. This is not a novel idea. Fine art connoisseurs and conservators have this experience when they explore a painter's unique handwork, traced in brush and other toolmarks on the painted surface. The challenge for the analyst is finding the places in a picture where their perceptions and interests seem to converge.

This is the point in analysis where words need to fail. In the interpretation of a product, the meaning is not the artist's, as a physical and social being existing at a single point in time; it is a narrative about the work that an interpreter creates *post hoc*, within a wider context of information in a society. Thus, Van Gogh is neither a troubled artist nor a man of his time, but rather a node in an endlessly shifting network of relations, contingent on the time and situation of interpretation. Only the destruction of the work ends its parthenogenesis.

In an outcome, the interpretation of meaning is more like the study of an exposed living floor in an archaeological excavation, where the clues are in the nature, distribution and relations of the artifacts exposed there. This too supports a narrative, but it is naturally grounded in the material evidence, and uses the material configuration to assess the impacts of forces or affordances in the surrounding environment during the production of the work.

If we were to assess Van Gogh's art from this perspective, the results might be disappointing, as we could not simply apply the putative knowledge about his psychological history to individual pictures. Indeed, there is a persistent myth that his *Wheatfield with Crows*, a painting he produced a short time before he died, was his final painting. Its visible turmoil, the dark skies, scudding clouds and roiling wheat, and the murder of crows swirling through the winds, suggested the advance of a storm, and this satisfied many of his contemporaries that it mirrored his apparent psychological state. It may also be viewed with less foreboding, however. As Bailey notes, "… along with its sinister aspects, it is also full of life – the ripening and swaying wheat, the soaring birds and the volatile weather" (2021, 112). None of the paintings that Van Gogh produced in the last days before he shot himself betray any hint of the darkness within.

Resistance to the lure of narratives clarifies the essential problem analysts and interpreters face as they move forward in a situational analysis. Whether the surroundings are bare, eroding earth or a rock outcrop or stone wall, such interaction requires the justification of the most basic of shared circumstances – the commensurability of perception – in this case, while standing and looking up today from the floor of a church at frescoes as the artist did in the 15th century.

Ecological perception and pictures

Overcoming the relativism induced by time, culture and subjectivity seems impossible, but turning to the open field of possibilities that display affords, we may draw on ecological perception, and the theory that the structure of the world is as we perceive it (Gibson 1966; 1978). This radical option is taken up infrequently in the study of art, as it does not fit well in an intellectual economy devoted to representations. To accept a perceptual ecology, however, it is necessary to reject representations as the basis of human action. Eleanor Gibson, who applied this approach in studies of children, presented the difference bluntly (1995, 13).

> A "self" that is a representation, or a concept of an executive made of mental stuff that directs the lower bodily orders, is a fiction. We are ourselves because we function as individuals who interact intentionally with other individuals and the world in which we live.

This epistemological turn is a tall order in any cultural pursuit. After more than two thousand years of scholarship and six hundred of humanism, assumptions about representations are so entrenched in the study of art and so consistent with ordinary understanding that even James Gibson had difficulty applying his theory to pictures.

The problem is the depiction. A picture has a material surface, an optical array that reflects light, but its depiction lacks most of the optical features of the lived-in world. In the journal *Leonardo*, Gibson concluded that a picture as a direct experience was impoverished: "mostly what is lost in a picture, drawing, painting, or photograph is the information that can only be extracted from the changing perspective structure of the ambient optic array of a moving observer" (1978, 234). He would call this elsewhere a frozen array.

He thus throws up an obstacle. To constitute a situation of production, a picture must be a physical environment in which the perceiver is active, and if this reality is denied, it seems difficult to use any form of ecology in its interpretation. Gibson's successors continue to work through the difficulties. Harry Heft, citing Susan Sontag (1977), regards the perception of pictures as "a special kind of visual experience ... characterized by a distancing of oneself from the subject represented, a holding of the world at arm's length" (Heft 2010, 28). Alternatively, Alan Costall looks outside psychology to anthropology and studies that regard meaning as implicit in social interaction, insisting that "The ecological approach needs to be worldly" (2017, 228).

The idea of relational meaning between an individual and an object gives new life to Gibson's frozen array. Yet, one problem in the analysis of pictures always brings up another. It might seem obvious that there is always a *reason* for a picture and it is something other than what it depicts, as Carl Knappet puts it, because a picture "signifies rather than acts" (2004, 43). The next question is inevitable: whose meaning is it? In a traditional dualistic approach, where meaning and materiality are categorically separated, the answer is "the representation". As Ingold has pointed out, however, this sets up an impossible condition, in which culture and its implicit ideas hover over the material world. His solution is to see an intellectual convergence of the two, in which meaning permeates cultural things (2000, 340).

This presents a difficult choice, however, in moving ahead with perceptual ecology in a study of situations. One option mediates between two worlds, one of them with no anchor in reality, and the other simply absorbs the whole into the totality of culture. If perception is a meaning relation, the problem is that nothing can exist outside its grasp – despite the notion that in direct perception, structure in the environment is always indifferently available to a perceiver.

As a result, this latter option, which seems so compatible with the pragmatic philosophies of Dewey and Mead and Gibson's ecological perception, fails on two accounts: under the dominion of culture and its intrinsic power relations, meanings exist at the scale of cultural and social processes and, thus, ordinary perceivers – individual people – have no freedom or visibility, practically or ontologically; and with meaning encompassing environment, studies open to ideas outside institutions and their governing constructs, and any sense of mutualism among individuals and environments is impossible.

Once again, it seems that pictures have a special nature, bound up in meaning – and this recalls the assumptions of the church fathers almost a thousand years ago. If a picture is no more than this, then there is little difference between Biblical exegesis and the study of the past, except that culture replaces the Bible as the frame of discussion.

The search for meaning should not forget its material source in the body and senses. Words and thoughts alone fail to account for pictures because most physical action never rises to awareness. For all creatures in motion, urges and imperatives intrude from all parts of experience, and so, unless one simply accepts the totality of humanism, the reasons behind human action and production remain elusive.

Affordances and invariants

In perceptual ecology, the problem of meaning in affordances distracts from another important aspect of Gibson's work that has some promise in integrating mind and matter.

Gibson's early work was on the problem of pattern perception; as the eye is always in motion in relation to the subject, and the optic array is continuously transformed, but in active perception, as an individual moves within an optic array, some features persist through movement.

> Perhaps the clue we are seeking lies in the invariant properties of such a continually changing retinal image. Only these properties would be capable of providing the stimulus basis for a stable and unchanging visual world.
>
> (Gibson 1950, 154)

When his later work inevitably came to the problem of pictures, he conceded that there was an intrinsic, structural relation between a picture and what it depicted, with structures "that uniquely specify their content" (Costall 1985, 25). Thus, whether the image of a tree was meant to be a tree or the godhead, it might be possible to agree that forests had a number of the objects depicted.

While the underlying processes involved in this interesting relation are as elusive as issues related to meaning, at least invariants enter into common life experiences. Costall once pondered how a picture works in a real-life situation, when someone waits to meet a stranger at a train station with only a photograph to go on (personal communication, September 7, 2019). As such dilemmas usually meet with success, there must be more to pictures than frozen arrays, and less than images in the brain; certainly fundamental perceptual processes related to face perception are involved (e.g., Leopold and Rhodes 2010). What he concluded from his wait at the station provides an essential insight into the grounds of situational analysis: the resemblance that invariants afford exists in "the forms of life they resource".

The role of common physical and environmental experience in social interaction provides a way forward in the ongoing puzzle about resemblance, both in the acts of an individual who picks out a face in the crowd from a picture and one who finds that the creature in a picture on a cave wall painted 45,000 years ago still occupies the forests and grasslands today – if it is possible to resist the ancient homocentricity inherent in cultural studies.

Across the golden divide, neurologists doing research on facial recognition have recently discovered that the brains of humans and monkeys have groups of neurons that fire when the subject looks at another's face; these areas are known as face patches (Tsao, Freiwald et al. 2006). While a neuron fired in a lab is a long way from a direct encounter in the wild, the direct stimulation of face patches by perceived features is consistent with how perceptual ecology describes the process of picking up visual information directly from the environment.

There is no reason to think that the restlessness of the eyes, body and brain when taking in faces or other parts of the perceived world is any less so in front of a flat picture surface. Indeed, it might be preferable, and closer to reality, to think that the actions of the artist produced the picture, that it emerged during the task, as opposed to something born as a representation.

The painters in Tuscany in Piero's time were clearly aware of the uncertainty of vision. Brunelleschi experimented with linear perspective, Alberti wrote his primer and Piero worked on the geometry of perception throughout his life to give them more understanding and control over the views of both artists and spectators. As with every attempt to rationalize visual perception, however, this experiment ultimately failed, because the building of any structure – physical or metaphorical – implies knowledge of the engineering first. Costall suggests instead that we "… think of any picture as a 'contraption', a more or less happy selection of formless invariants" (1993, 336).

From this we may conclude that a picture is not some imagined wholeness or logical unity, which suits the idea of a representation, or a relation that valorizes concepts of meaning and culture, but rather the deeper and more enduring convergence implicated here in display. This is not a primitive or prehistoric condition, which implies that humans in the present are the center of the world; it simply aligns to human physiological and perceptual behaviours that are essential to culture but do not address culture at all.

In a recent discussion of how to let go of traditional concepts of linguistic relativism and look more clearly at commonalities in speech situations, Zinken and Costall suggest that "close attention to action-in-a-situation helps us ground empirical work on cross-cultural diversity in an appreciation of the invariances that make culture-specific elements of practice meaningful" (2019, 491). Fundamental in any situated human action is the cascade of active perceptions that attends ordinary daily life. While the diversity of picture contents are endless, the medium carries situational correspondences in its sensory display that are similarly visible across cultures.

Thinking back to rock art in the Canadian Shield, in the slow drift of the canoe to a rock face, it is first the redness of the paint and then the conspicuous forms that capture the eye and anticipation against the flows of dull, grey granite, long before the pictures present themselves to interpretation.

Display as communication

The independence of invariants from the specificities of meaning suggests that we should pause in this endless search and consider the power of the perceived object. As Marshall McLuhan observed, "it is only too typical that the 'content' of any medium blinds us to the character of the medium" (1964, 9).

McLuhan's ideas about the transmission of information provide more insight into how the physical condition of pictures may influence their perception, in particular his distinction between the stable carrier of information, the medium, and its ephemeral, varied messages. He paints with a broad brush, insisting that the medium holds the real power in communication. While messages reflect the endless diversity of people and societies, the medium "shapes and controls the scale and form of human association and action" (1964, 9), and it does so without reference to its contents. Today, the communication machine of choice is the smartphone, a carrier of brands and contents as various as there are users, but the great influencer is the device itself, which has changed the habits of individuals and the dynamics of social relations at a global scale.

Looking at a picture is more complex, however, because unlike other arts or information machines, it acts as both medium and message. The messages seem clear, if elusive. They are the putative content, imagined by a conjunction of artist and spectator, whether at the time or any time in the future. The medium of a picture, however, is deceptive. For McLuhan, concerned with communication at the scale of societies, the carrier of information has an underlying order. Noting that people always ask what a picture is about, he writes: "they never thought to ask what a melody was about, nor what a house or a dress was *about*" because for people these still carry "some sense of the whole pattern, of form and function as a unity" (1964, 5).

Here we depart from McLuhan's concerns, but only in its modernist implications. From the ground, the unformed, *underlying* unity in all pictures is not ideological or ideational notions drawn from the product, but rather relates to the expressive act. If communication is the goal, natural selection and common experience suggest that it is most effective if it draws attention through salience. The commensurability of the senses is why neither the painters and carvers who worked over thousands of years in Lascaux cave, nor the boys who first discovered the paintings, needed instruction in art or art history to see the images.

This salience is no different than emphatic utterances in speech or the dynamics of sensory attractants tested and persistent through natural selection. Of course, as a message, a picture may inspire a universe of ideas, compared to a bowerbird's mating den or a stop sign, but ideas only obscure our physical dependencies. When an artist fashions a visible display, whatever the rationale, it joins them with ecological legacies which extend out of humanist representations into wider forms of life.

In this way, picture perception resembles perception in an outside environment. On the ground, at the scale and sometimes in the setting that the artist also occupied, the dynamic relations implied in the salient features of display are consistent with what perceptual ecology contends. Individual people actively move or search through their surroundings and pick up structured information in the diverse optical arrays available to them – without the *need* for the intrusion of authorities. This is because the knowledge is in the cascades of active situations and relations and the unformed invariants in the senses that constitute the already experienced world. Facial invariants help people recognize strangers from pictures, just as the realistic faces in Giotto's or Piero's frescoes make legendary figures seem familiar and, so, more accessible to those who believe in them.

The direct experience of an artist's work therefore provides another way to learn more about what a picture communicates. In the artist's patterns of salience we have affordances to interaction associated directly with meaning, and even if it is impossible to do more than speculate about the artist's intent, having some tangible connection through physical experience is better than having none at all.

The shock of salience

To explore the features that an artist creates to draw attention and to communicate is like observing strangers in a distant conversation. We may recognize them, see their expressions and movements as they talk, and guess what they are talking about through their visible actions, but the information they share is elusive. At least in this circumstance, however, we have some empirical grounding as we begin to inquire about what went on.

What attracts attention in the perceived environment, whether a small group of people, a face or a picture, is a salient feature. Gibson observes this phenomenon in ordinary perception:

> Only in an unfamiliar environment or a problem situation do we become fully aware of the literal visual world. One has to pause and look in order to see it.
>
> (Gibson 1950, 21–22)

McLuhan also sees pictures in this way, as anti-environments, which, compared to the difficulties with perceiving order in outside environments, "provide means of direct attention and enable us to see and understand more clearly" (McLuhan and Fiore 1996, 69).

And Michael Polyani writes that "By means of its artificial framework, that is sharply incompatible with its subject, a work of art takes us into an experience beyond the realm both of nature and of practical affairs …" (Polanyi 1970, 230).

Underlying these ideas is the tacit assumption that a picture represents something to be socially defined, whether in models of social processes or in higher level gestalts, associated with the mind. If we apply the ideas of dynamic interaction among senses and ideas in perceptual ecology, however, we have a way to approach the problem of meaning while still looking at the picture – by focusing on the shock of salience and its distribution in an environment.

In the uncontrolled flows of information in a locality, salience may be an irruption intended to draw attention to some message or other content. In a distant conversation, it can be as simple as guessing what is going on by the tones and what we take to be emotions. But it can also be the way they appear, their deportment and gestures. And this takes us more broadly into sensory perception.

Consider the trappings of ritual. Most often, a shaman or priest, working in an enclosure that keeps the profane world out, attempts to shock participants out of their ordinary state with mysteries conveyed through provocative actions and equipment – a vast array of processions, rituals, patterns of movement, candles, censors, bells, chants, music and song, painted or carved images or ornamentation and other paraphernalia. And, of course, salience extends into the streets, on the bodies of individuals. Striking costume and adornment, tattoos, perfumes and other forms of sensory display may be rationalized in many ways, but behind them all is the tacit sense that they will attract attention and provoke response.

This joining of sensory energy and message is also the essence of the elaborate spectacles and monumental works in every political leader's toolkit, where salience becomes a direct way of attracting, directing and educating the attention of masses to the bearer of the messages. As salience is joined to the power of attraction, it may be used in an archaeological context as a means to assess the conditions that provoked its use.

A brief example from Eighteenth Dynasty Egypt (1549–1292 BCE) shows how the ebbs and flows of political power may intrude into daily life, appearing in the depiction of social relationships in tomb painting (Molyneaux 1997). The attention was on the actual work of draftsmen and painters in the dark and cramped quarters of individual tombs and the fictive settings of their depictions. The question was whether or not the circumstances in Egyptian society with respect to notions of power induced variations in how these artists depicted the tomb owner and others in the typical offering scenes of the private tombs of high-status commoners in this period.

The most striking effects proved to be a subtle aspect of visibility – the relative heights of interacting figures. In offering scenes, the tomb owner was always slightly shorter than the god; and where he was depicted or in the form of a seated statue with his wife, he was always slightly taller. The sizes of other figures, however, were highly variable. When the tomb owner was pictured offering to or receiving gifts from the king, their relative heights varied substantially over the time period.

Egypt became increasingly powerful and prosperous during this Dynasty, and varied fortunes of its kings may have influenced the depiction of relative equity of the interaction between the two as represented in the offering scene. The reasonable assumption was that the visible inequality in the relationship in the picture varied according to the need for ideological reinforcement of royal power in the society itself.

The visible difference between the tomb owner and king in the tomb paintings gradually increased through the dynasty until the reign of Amenhotep IV/Akhenaten, when the king and queen, Akhenaten and Nefertiti, towered mightily over the tiny tomb owners and other supplicants. Following Akhenaten's death, this style lingered briefly, but then quickly returned to the ancient conventions in the last ruler of the Dynasty, the usurper Horemheb (1319–1292 BCE), a commander of the army under Tutankhamun.

The result suggested two distinct forms of relationship displays. One was constant and, therefore, conventional – the relations between tomb owner and god, and tomb owner and spouse. The other was varied and political, related to the expression of the relative power and status in the relationship – fictive or real.

To account for Akhenaten's exaggerated displays, we turn to circumstances outside the tombs. When Amenhotep IV chose to take more political power for himself, the response was more than a slow intensification of power. Monumentality was the norm in Egypt, and in his new persona as Akhenaten, his self-aggrandizements were typical. In the tombs, however, the difference was so stark compared to earlier reigns that a new dynamic might have come into play – the need for increased ideological reinforcement. Consider his situation. He shifted the focus of religion to a single god, the Aten, and moved the capital to a new site in Middle Egypt which he called Akhetaten, "Horizon of the Aten", to house his ambitions. The transplanted population was probably disordered and restive and, under these conditions, the king may have even more dramatically reinforced the representation of his power. The exaggerated salience of Akhenaten's displays in the tombs of the new city seems to bear this out and, after his death, the depiction of a tomb owner as only slightly smaller than Horemheb suggests its collapse.

The work in Egypt suggests that depicted social situations, however fictive they might be, are as sensitive to flux in information flows as actual ones. Treating pictures in general as depictions within situations of perception, action and production thus creates the potential for new research in cultural life that seeks associations for observed anomalies in presentation in the social and material surroundings, as opposed to applying data indifferently from models and representations defined through discourse.

Assessing salience for its irruptive qualities provides an information variable that ranges between two practical extremes. There will be times when there is too much information – a condition of sensory or ideological noise. One has to shout above the din to be heard. This was life then in the new royal city of Akhetaten, or now in Las Vegas, where casinos compete in a forest of neon signs (Venturi et al. 1972). In other situations, there is not enough, when it is necessary to rouse someone into action from habitual life. This is most clear when art becomes a tool to engage the masses.

In 1917, Shklovsky (1965, 31) complained that:

> we apprehend objects only as shapes with imprecise extensions; we do not see them in their entirety but rather recognize them by their main characteristics. We see the object as though it were enveloped in a sack.

His solution was defamiliarization – distorting conventional appearances in art to shock the viewer into action, so that they would look again with clarity at the surrounding world, "to make one feel things, to make the stone stony" (1965, 32).

What Shklovsky, Gibson, McLuhan and Polyani understood about the shock of pictures is part of the same legacy of display that emerged long ago in the Pleistocene. When hominins first exploited colourful pigments and created visible forms in a variety of media, they introduced frissons into ordinary life in an entirely traditional way, as a means of attracting attention to themselves. Tens of millennia later, graffiti art has replaced rock art, but the medium and its sensory dynamics remain the same.

In the basilica of San Francesco, the friars ostensibly planned its imagery for its didactic effects, and so the larger-than-life frescoes have a salience like that of a monument that dominates the horizons and organizes the life around it. Gazing up at them, however, it is hard to imagine that any scheme of ideas could overwhelm the sensory richness of the encounter and the curiosity about the people and the human events that the images depict.

The question going forward is therefore about specific intent – whether Piero adjusted the visual and narrative flows of his pictures for his own purposes, outside the ordinary intentions of the friars and patrons and the conventional expectations within Tuscan society at the time.

Interventions in canonical art

During the 15th century, a painter faced a more complicated situation than conforming to the requirements of the clergy. With increasing prosperity in the city-states of northern Italy, wealthy families commanded more power and influence, and while religion maintained its visible authority and continued to be the main venue for the public display of art, there were new opportunities for patrons. Wealthy citizens could gain recognition for their good works and piety, and so they commissioned artists to find solutions that did no violence to social mores.

From an artist's point of view, it must have been a challenge to manage the attention of spectators to matters outside the canonical displays. Giotto faced this problem in planning and decorating the Scrovegni Chapel in Padua, which he completed ca. 1305. He was clearly more than a painter of the biblical scenes he was obliged to create. He painted faces of people and surroundings he saw in life, which takes the viewer's imagination out of the picture, the church and into the streets. And he considered the church itself as an environment with an audience. This seems clear in the decoration of the triumphal arch. The Scrovegni made their fortunes in usury, and Giotto's subject here acknowledges the sin (Stubblebine 1969, 85).

The realism of faces extended to the gaze. Early in the 16th century, Raphael (1483–1520) painted an altarpiece dedicated to St. Cecilia. While the composition ostensibly centers on Cecilia, looking up to the heavens in rapture, it is difficult to avoid the gaze of Mary Magdalene, who looks straight out towards viewers. The directness of her stare seems to have had little iconographic potential in art history, given the brief reference, for example, in Mossakowski's study (1968). Indeed, he repeats Muntz's (1882) rather bland idea that "St. Mary Magdalene turns toward the spectator, as if to invite him to share her feelings" (1882, 673).

In a study of women and patronage, however, King (1998) finds much more in this glance. She emphasizes that it was a woman, Elena Duglioli, a member of an aristocratic Bolognese family, who commissioned the altarpiece. Duglioli led a spiritual life, devoted herself to charity and other good works, and was an example to many people at the time. She was especially devoted to St. Cecilia, and received relics of her from Bishop Francesco Alidosi, who was Cardinal of St. Cecilia and the Pope's legate at Bologna (Mossakowski 1968, 16). Patronesses had more difficulties managing their displays than men, as they had to bear more spiritual and social constraints, but they found an opportunity in managing the gaze. As King suggests, Elena Duglioli may have directly or by example inspired the Magdalene's gaze as if it was her own. We might even turn Muntz's observation around altogether, so that when a viewer meets Magdalena eye-to-eye, the picture frame dissolves and she comes out into the world in which Elena lived, where all her devotion was manifest. This seems a better fate than leaving her in the picture under the authority of a patriarchal belief system or captured in Raphael's oeuvre and the ideals of the higher Renaissance.

At this point we may draw explicitly from Goffman's (1961) idea of secondary adjustments, which he conceived in relation to total institutions that attempt to manage all individual behaviour (he studied prisons and asylums, but any totalizing ideology will do). He observed that not all individuals adjust their behaviour to the norm; some choose to make secondary adjustments, working around or defying the imposed order to gain some small sense of power or freedom.

Among painters and scholars in the 15th century in Tuscany, there was a saying that a painter paints himself (Zollner 1992), and that some of the seemingly personal references might be a kind of automimesis. The gaze of St. Magdalene, however, suggests the opposite, that some artists made secondary adjustments under a strict visual convention to show their independence from dominant narratives or ideologies in their own times.

With these situational dynamics in mind, it is possible that the puzzle of Piero della Francesca's fresco cycle in San Francesco is a secondary adjustment to life in Arezzo. The local focus is crucial here, and practical. Piero worked with his assistants to present ideas urged, compelled or inspired by the diverse audience that included religious and civic authorities, patrons, congregations, fellow artists and friends. No-one can be certain what he meant in the work, but we are more certain about how he meant it, as he displayed his choices on the wall for all to see. Perhaps some of the answers to their curious organization are also still there, visible from the church floor.

Before the frescoes

The commune of Arezzo was an important city in Tuscany during the medieval period. It occupied a strategic location near the headwaters of two major rivers, the Arno and the Tiber, as well as the overland routes from Rome through the Apennines to the east coast. The commune asserted itself in the various territorial wars and quarrels or the control of transport and trade, especially with its larger regional rival, Florence, until June 11, 1289, when the Florentine forces overwhelmed the Aretines at the Battle of Campaldino (Oerter 1968). A weakened Arezzo was then caught up in the struggle between three rival bishops with claims to the papacy, which carried on between 1378 and 1417. The city was first occupied in 1380 by Charles of Durazzo, a member of the Neapolitan Angevin royal family, supported by the Roman Pope Urban VI, and then in 1384 by the mercenary Enguerrand de Coucy, under Urban's rival Louis of Anjou, the brother of the King of France. Arezzo's independence came to an ignominious end when Florence obtained the town from de Coucy ("bought off by a large loan" (Holmes 1989, 293)) and it was quietly absorbed into Florentine territory.

By Piero's time, barely 50 years later, Arezzo was a provincial town under Florentine rule, but memories of its ancient independence and more recent fate were probably still fresh when Piero and his workshop began to plaster and paint. Unfortunately, the situation in Arezzo at the time consists mainly of footnotes to matters in Florence. This city not only dominated Tuscan life at the time, but also the assessment of it, as it became the center of the emerging Renaissance.

The enthusiasm for the historical representations supporting the idea of the Renaissance is problematic for the study of any single artist at the time, but it was especially acute for painters because of a singular act of creation by a member of a wealthy Florentine family, Leon Battista Alberti (1406–1472). In 1435, he completed a small treatise on painting, which he circulated in two versions: one in Tuscan, *Della Pittura*; and one in Latin, *De Pictura* (Alberti 2013).

Alberti began his Tuscan version with a dedication to Filippo Brunelleschi (1377–1446), the architect and engineer who designed the dome of Florence Cathedral and used classical optics and geometry to create a novel form of linear perspective for drawings. This dedication was quite apt because he begins his text with a primer on applied linear perspective, and then moves on to write about the higher nature and significance of art and artists – all of it written in the idiom of the classical thought and style in which he was educated.

As this treatise began to circulate less than 20 years before Piero took up his brushes in Arezzo, it is difficult to assess its influence. Piero worked in the same intellectual climate, as he used the device for some effects in the Legend, but he came from a modest family, lived in a different social milieu, had different interests and, eventually, was forgotten for several centuries. His great treatise on the mathematics of linear perspective, *De prospectiva pingendi*, or "On the perspective to paint", which Vasari cited in the next century, does not exist in an English translation.

What is most apparent in the background research and interpretation of Tuscany in the 15th century is the Florentine domination of scholarly output and its subsequent treatment as a primary source of humanist thought. It transformed the idea of art at the time from simple depictions of beliefs and their symbols to a portentous, intellectual act. Edgerton (2006), for example, characterizes the frame that Alberti uses to look through and develop his perspective relations on the world as important in "redirecting the purpose of perspective art away from revealing God's divine order as reflected on earth, to a more secular physical reality viewed directly in relation to human moral order" (2006, 151).

Alberti was complicit in this. His window was not meant to open into a real world with an ordinary nature. The task of the artist was to create a refined nature, suitable to the higher purposes Alberti proposed for artists like himself.

Given humanism's preoccupations, and given that artifacts only make sense in the context of their creation or use, it seems prudent to stay as close as possible to the visible world that Piero displayed on the walls of the basilica of San Francesco (see Field 1997). Here, we are at least certain of the physical environment within which he worked in the chancel. It is a relatively small space with high walls behind the altar and screen of the choir, lit only by the window at the back. Fortunately, Bicci had finished the painting in the difficult architecture of ceiling vaults, leaving Piero to work on the two side walls and around the window. We can also be sure that every day he worked on this task, he entered the church from the steep, cobbled streets of Arezzo, walked through the choir and entered the chancel. There, he and his assistants worked from scaffolds to depict events in the two accounts that make up the legend of the true cross in Varagine's text, displaying all of them above eye level on the three bare walls.

As a painter, Piero would have inhabited the work like a builder who lives in a shelter while building it. In these acts, he did not represent ideas, but rather faced the wall and planned, sketched, applied pigment to wet plaster and pondered the results as the work progressed. And, just as the legend of the true cross wound through spiritual and actual history, he must have navigated through the wishes and demands of the friars and the Bacci family and all this within his own daily life in Arezzo and beyond.

Until the late 20th century, spectators gazed up from the floor at the pictures illuminated by the shifting intensity and colour of natural light. An old Baedeker travel guide noted: "best light about midday and in the evening" (Baedeker 1909, 52). Since then, however, the frescoes have taken on an appearance that they never had in nature: deteriorated images have been restored, the church has electric light and, most radically, the frescoes are available remotely in closeup photographs, taken not at the steep views from the floor, but from Piero's unique perspective on the scaffold.

While the digital records of the images are stunning in their visibility and detail, the view from the floor is crucial to understanding Piero's work. It was one that Piero knew well as a spectator, and included the people crucial to the success of his project. When the task and its ordinary and privileged perspectives are considered together, evidence of Piero's interventions emerges.[1]

Piero's interventions

Jacopo da Varagine ([1483] 1900) describes the most ancient history thus:

> The holy cross was found two hundred years after the resurrection of our Lord. It is read in the gospel of Nicodemus that, when Adam waxed sick, Seth his son went to the gate of Paradise terrestrial for to get the oil of mercy for to anoint withal his father's body. Then appeared to him S. Michael the angel, and said to him: Travail not thou in vain for this oil, for thou mayst not have it till five thousand and five hundred years be past, how be it that from Adam unto the passion of our Lord were but five thousand one hundred and thirty three years. In another place it is read that the angel brought him a branch, and commanded him to plant it the Mount of Lebanon. Yet find we in another place that he gave to him of the tree that Adam ate of, and said to him that when that bare fruit he should be nourished and all whole. When Seth came again he found his father dead and planted this tree upon his grave, and it endured there unto the time of Solomon. And because he saw that it was fair, he did do hew it down and set it in his house named Saltus.

Piero began this story high on the right side wall in the lunette with three scenes depicting the lingering and death of Adam set against the background of a great tree in the Garden of Eden.

Beginning on the right, an aged Adam sits on the ground, tended by Eve, his son Seth and a young man and woman looking on.

Next is a tiny scene, set in the background, depicting Adam's son Seth asking the Archangel Michael for God's help. It acts as a kind of punctuation between Adam's failing and his death, just visible enough to maintain the logic of the narrative. In the rest of the space the family push forward in mourning over Adam's supine and shrouded remains.

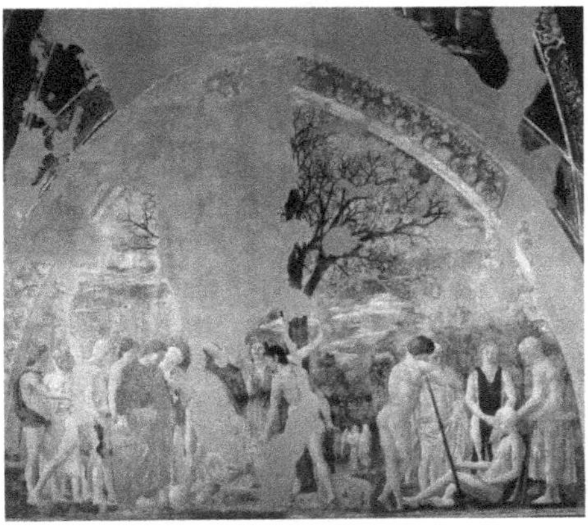

Figure 7.1 The Adamite Lunette

From the ground, it is difficult to pick up finer details, but in each group, Piero draws the eye into the scene by emphasizing a single figure. At Adam's failing, it is the young woman. She stands behind Adam, clad in a black animal skin draped over her shoulders, and faces outwards towards the viewer. In the funeral scene, the woman who stood by the failing Adam again stands in the center of the crowd of mourners – and in the center of the whole lunette. She now wears a white gown. As she mourns over Adam's body, her face contorted with grief, she stretches her arms outwards, an act that resonates in the limbs of the great tree behind her. And at the left of the scene, among the mourners who strain forwards to see Adam, another young woman catches the eye. She stands with a young man, slightly separated from the rest of the family. Instead of looking towards Adam, she gazes outwards, her accentuated dark eyes open wide, towards her companion.

From the ground, it is quite clear that Piero has used these three female figures to organize the lunette and manage the attention of the audience across its breadth. They are spaced equally, anchoring the visual flow across the composition, and, conspicuously, they each seem to address the audience below.

Only by using the virtual scaffold can we see Piero's detailed work, and this gives us more clues about his intentions. The woman in the couple is significant in the structure, but seems peripheral to the scene by her position and the direction of her attention. Close up, she seems to gaze more directly towards her companion, and we now see that he returns the gaze, and his slightly parted lips suggest that he whispers to her. If this visible separation was not enough, the jumble of feet visible below the hem of her gown reveals that another mourner is obscured behind the couple – and so they are indeed within the group. By introducing this uncertainty in where the couple stand – in place and society – Piero accentuates the anomaly in their behaviour.

Why Piero painted this small vignette and created an emotional frisson within the group at the funeral is a mystery, and several interpretations exist that one might choose from to explain it, but it clearly draws attention away from the story of the true cross and its unfolding. It is as if he steps outside the narrative to make a subtle comment to some in his audience.

A small disturbance in the narrative flow also occurs in the next part of the story, in the register below. After many years had passed, King Solomon finds the tree and cuts it down to use as a bridge at his palace, unaware of its significance, until a visit by the Queen of Sheba. As the Queen arrives at the bridge she is seized with a vision that a timber on this bridge was to be the holy cross. At this point, the story shifts from Biblical history to the legend, as the Bible records their meeting without mentioning this, or the events to follow (see 1 Kings 10 KJV).

Piero paints both episodes in the register, suitably focusing on Sheba and Solomon, but he also ensures that restless eyes in his audience look elsewhere. The scene of Sheba's veneration of the sacred wood combines two events. Sheba, supported by an entourage of ladies in waiting and a servant, kneels in

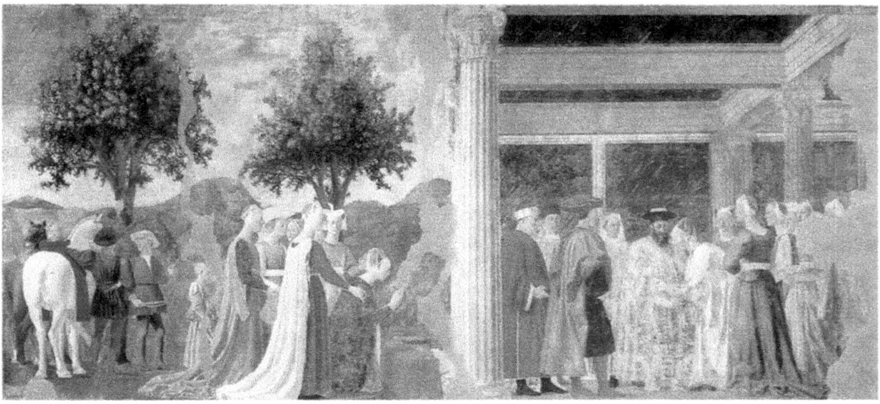

Figure 7.2 The Meeting of Solomon and Sheba

front of the sacred wood and prays; and at the back of the procession, a pair of courtiers wait with their horses. As Sheba prays, two of her attendants look out of the scene in different directions, distracted. The intent to disrupt is clear: one woman is at Sheba's left side, behind her in the picture plane, and, like the women in the Adamite lunette, she looks out of the scene towards the audience below while gesturing towards Sheba with her left hand, half hidden behind Sheba's head. The diminutive woman wearing a distinctive cone-shaped hat, who watches attentively at the end of the procession, must be a servant attending Sheba. Like the tiny scene of Seth and the archangel, the space she occupies yields to the larger space with the men and their horses. This is not the background of the devotion; this group takes up almost half the panel – yet the men have no interest at all in Sheba's veneration. They hold the reins of their mounts to steady them and they converse – not likely about Sheba and the sacred wood – while a brown steed at the left edge of the scene shows the whites of its eyes and bares its teeth, clearly agitated. Like the young couple above, it distracts from the main events. Ordinary life with its scatter of perversities goes on, it seems, and perhaps to some who see it from below, it is a counterpoint to Sheba's prayer.

Piero's distractions continue into the meeting of Solomon and Sheba in the temple. One of the men in Solomon's retinue prefers to look out into the chancel, with a face and riveting gaze that is clearly visible from the floor. In the city of Arezzo today, this is presented as the face of Piero himself.

We will never know Piero's reasons for creating these small irruptions in the visual flow, but he seems to have asserted his own authority over the narrative from the outset. The Legend is about the true cross, yet, Seth's intervention with the archangel, when he receives the vital branch, is merely a pictorial device. Piero clearly chose human drama over angels. By managing attention between views from the floor and from the scaffold, he gives the visual reading favoured by the church fathers some of its due, but not too much. The restless gaze of some of his characters draws attention out of the story, encourages interaction with viewers and, in this way, gives space and presence to ordinary life (Baxandall 1972, 75–76).

After introducing the cross before the meeting of Solomon and Sheba, he moves to the illustration of the Legend, beginning with Solomon's burial of the holy wood "dolven deep in the ground" (Varagine 1900, 170), which he places on the adjacent back wall – but then he strays from the linear authorities of speech accounts and writings, and the existing fresco cycles of Giotto and Gaddi, as if such authority did not exist.

Piero's scheme for the rest of the Legend has provoked much comment over the centuries, but perhaps he had something more in mind than fidelity to an imposed narrative and more than a single audience.

Finding the audience

Thinking about the audience was not an invention of the emerging Renaissance. The way individual people saw and used images was a central problem in the church from its beginnings. The church fathers were not artists, but they understood the problem with pictures, and never abandoned the belief that clergy and congregations might draw from them what their faith demanded.

What seems to change is the artist's perception of the audience. In paintings that explicitly serve ideology, from images of the sacred to more recent notions of pictures as a kind of visual language, the idea is more important than any particular depiction of it. This would have suited the more literate clergy, who had their manuscripts, and, indeed, Giotto's frescoes in the Scrovegni Chapel convey this sense of being enveloped in one. However, when painters begin to visibly assert their authority, as Alberti does in 1435 with his artist's view of theories, methods and the rewards of painting, they had to attend to the audience who would see these works from the ground. For Piero and other artists following Alberti the need to balance competing influences of religious and secular authority and society continued to increase, and these eyes mattered, spiritually, socially and economically.

Such pragmatism may account for the renewed interest in optics. In theory, this concern for perception was an aspect of Renaissance ideals, but in practice, it seemed to be an ideal way of drawing viewers to look at specific relationships within a larger scene. In this regard, Piero shows his mature understanding of situations. He did not rush to apply a linear perspective. Looking from the chancel floor, his treatments seem rather academic, traced out most visibly in background architecture. Instead, he takes advantage of the cramped space and creates broad, flat panoramas of people, things and events that flow across the wall. Without the awkward device of a single viewpoint, the scenes appeal frontally to each spectator, no matter where they are standing in the space. And he chooses his moments across the wall for drawing the audience in and out, here and there creating the illusion of depth through the figures who gaze outwards or are set back in the distance. All this breaks the illusion that what appears on the wall is meant to reflect the greater authority of a spiritual world.

This more pragmatic treatment is most obvious in the least subtle of Piero's placings of the Legend's episodes. He spreads its two great battles across the walls of the bottom register, just above the heads of the audience, plunging them into the middle of the action.

The right wall depicts the legendary victory of Constantine over Maxentius, his chief rival for control over the Roman Empire, on October 28, 312, near the Milvian Bridge over the Tiber in the outskirts of Rome. Maxentius and his forces retreated across the Tiber, and Maxentius is said to have drowned trying to swim across. For those who know the Legend, the story begins by the rear window, where Piero depicts Constantine in his tent, dreaming of an angel and the sign of the cross which would give him victory in battle.

The other wall represents another crucial episode in the wars to defend Christianity, when the Emperor Heraclius marched through Persian territory and confronted the Persian king Chosroes near Nineveh in early December, 627. The forces of Christianity were victorious, and Heraclius retrieved the pieces of the true cross that Chosroes had taken and he eventually returned the cross to Jerusalem.

This final event is high above in the lunette on the same wall, the end of Piero's story.

It is difficult not to yield to the obviousness of this visual effect. Modern spectators may be inured to the linearity of narratives in books, moving images and digital pages. Until the late 19th century, however, this surround must have been stunning indeed. As Vasari (2012, 20–21) relates, "he represented most successfully in a battle fear, animosity, dexterity, vehemence, and all the other emotions that can be imagined in men who are fighting, and likewise all the incidents of battle, together with an almost incredible carnage, what with the wounded, the fallen, and the dead".

If we look at these two scenes as a moving wall, rather than a wall of meaning, its sensory drama appeals to more than a single audience. Someone familiar with the story will see the flow of battle move from the triumph of Constantine and Christianity to its defense, when Heraclius defeats the Persians. To others, not schooled in the history or iconography, and probably to all, visibly and emotionally, these walls present a timeless theatre of battle with its own internal logic. The action begins on the right wall with a march to battle, a melee of cavalry and hand-to-hand fighting and a chase and, on the left, the heat of the battle and the execution of an opposing leader.

Piero's emphasis on a visceral reaction is clear in his choice of superimposed planes to render the action. As anyone who has been caught in a surging crowd knows, there is no single point of view, no opportunity to draw back to and observe. This is especially clear in the battle with the Persians, as the chaos of the fighting spills towards the viewers. In the bloodied foreground, on the left, a decapitated head lies at the feet of its body, near the center a soldier lifts up his adversary by the hair after thrusting a sword in his

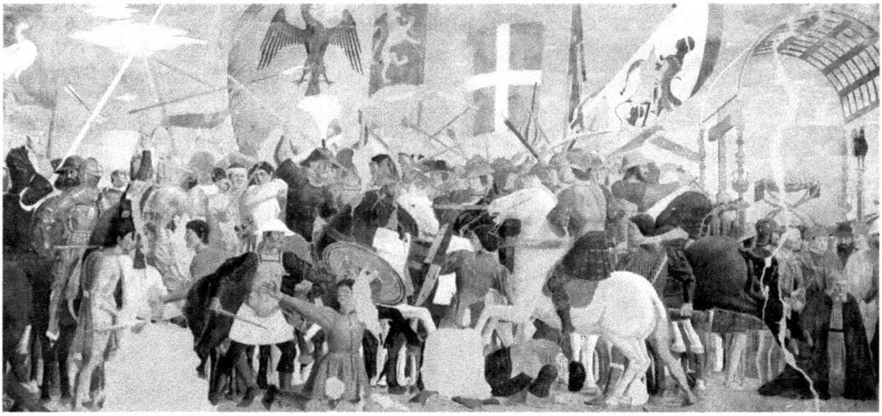

Figure 7.3 The Battle of Heraclius and Chosroes

throat, about to cut off his head, and to the right a horse charges into the fray from the space in the chancel.

It thus seems doubtful that Piero would choose to place the most exciting scenes nearest the spectators because they fit into some grand plan for the Legend. His various interventions suggest that he never saw the legend and its narrative as sacrosanct. Instead, he may have responded to more immediate political and social circumstances as he worked, from the view of the friars, his patrons or himself.

Religious politics would always loom in the background of a church, and Florence was the venue for an ecumenical council, active between 1438 and 1445, that attempted to bring the Byzantine Church on Christianity's eastern frontiers back into the fold. At issue was the rising power of the Ottoman Turks, carrying Islam, who pushed westward against Europe's eastern frontiers. More to the point, however, was the situation in Arezzo, where the population might have been much more sensitive to threats from outside, considering the Aretine's catastrophic battle with the Florentines in 1289 and the eventual loss of independence less than 100 years before Piero began his work.

It is also possible that the decision to surround his audience with battles was not part of the original plan at all, but happened in response to an immediate crisis. The beginning of Piero's painting and the sequence of episodes is not yet known, but if Piero worked from the top of the walls down, he may not have sketched out the battle scenes when news came that Constantinople had fallen to the Ottomans on May 29, 1453. The one ship that escaped arrived in Venice on June 29, 1453 (Setton 1978, 138), and the news spread quickly down the peninsula.

This calamity might have convinced the friars, the patrons or Piero himself that reinforcing Christian morale with stories of victories in battle was more important than fidelity to a legend.

The window wall

If Piero positioned the battles as a response to religious concerns, or even simply as an antidote to Varagine's dour and academic narrative, these scenes contrast greatly with the rest. Piero's other figures and landscapes have a lambent stillness that seems to suspend them in space and time. Yet, this calm only magnifies his frissons, from the small glances in the remote spiritual heights to the tangible chaos of the floor. Surely, he intended these to break up the visual flow and shock viewers out of the story to see *his* narrative. Some details, such as the lover's glances at Adam's funeral, may have been personal comments that he hid in plain sight, more visible from the scaffold than the floor. This should not be surprising. His spectator's sense and sensibility would have developed in the months and years he looked up from below at what he had made. Indeed, Vasari (2011, 21) provides grounds for seeing Piero's distance from the Legend in his comments about portraits of living people he supposedly included – doubtlessly comments Vasari picked up from talk in Arezzo:

> For this work, therefore, he well deserved to be richly rewarded by Luigi Bacci, whom he portrayed there in the scene of the beheading of a King, together with Carlo and others of his brothers and many Aretines who were then distinguished in letters; and to be loved and revered ever afterwards, as he was, in that city, which he had made so illustrious with his works.

If Piero inserted portraits of his patrons and other local luminaries in the Legend or, at least, painted faces one might see among the clergy and laity in the church, this might help account for his inclusion of the Annunciation in the chancel, painted opposite to the Dream of Constantine in the bottom register of the window wall. While some analysts have gone to great lengths to work this scene into the logic of the Legend, perhaps its placement here had more to do with the audience than the story.

The friars and other elites with the authority to walk behind the altar and linger in the chancel had the privilege of seeing the frescoes up close. As the most sacred part of the church, this space was visibly separate from the choir and nave and, during medieval times, partly obscured by altars and screens. Together, the architecture and furniture created perceptual and intellectual boundaries that matched the hierarchies of power and privilege in the society (Jung 2006).

The view is from the back of the chapel inside the altar, looking out towards the nave, where the congregation peers in. Women and the fortunate members of the congregation in the middle of the nave, who were all generally excluded from the more sacred areas, watched from the door.

Fortunately for the basilica of San Francesco, a detailed floor plan of the church as it would have been in the 15th century has survived, made by the architect Fra Giovanni da Pistoia(?) in the 14th century (Cooper 2014, 49). It shows a sequence of screens dividing the space and the audiences for Piero's art.

This plan shows that we have several stages of view to consider. In the chancel one is immersed, and the issues are vertical – picking out the details and meeting the gazes. Looking from the choir, however, and moving from in front of the altar towards what would have been the nave screen, the view of the frescoes on the side walls diminishes quickly, until they are too distant and at too severe an angle to be seen and, except from an opening into the nave at the screen, not visible at all. Thus, for choristers and clergy in front of the altar, there is no Legend sequence, only the back wall, its great leaded window surrounded by six frescoes – half scenes from the Legend, and half not.

It is possible, however, that the window wall was Piero's version of a Legend for this larger congregation. It is hard to imagine that he did not ponder these different perspectives as he walked through the choir or the nave each day to get to his paintings. Seeing as an ordinary man, not a friar, he would have noticed that the window wall dominated this larger space, and it may have inspired him to create a second version of the story.

In this more limited reading of the Legend, the top rank has two male figures presiding at the vaults. If they are the two prophets, Isaiah and Jeremiah, a cursory reading of the Bible can easily produce prophesies that relate to the troubling situation of Christianity in the 1450s: Jeremiah predicts disaster from the north (Jeremiah 1:14–16 KJV); and Isaiah (Isaiah 9:19 KJV) gives a call to action, filled with judgment and hope.

Descending, the middle register pictures two key events in the history of the cross. The scene at the right side of the window depicts the burying of the wood on Solomon's orders, because Sheba said "the Saviour of all the world

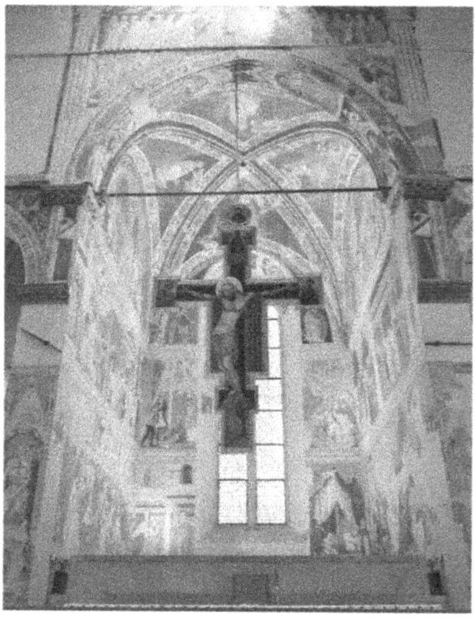

Figure 7.4 The Chancel Window Wall

should be hanged thereon, by whom the realm of the Jews shall be defaced and cease" (Varagine 1900, 170). And the left side depicts a man being hauled up from a dry well by a rope and pulley, and by the hair. The Legend recounts that thousands of years after Solomon buried the wood, and after the Emperor Constantine's victory in the name of Christianity over his rival, Maxentius, Constantine's mother Helena found the one man who knew its location – Judas, a Jew. He refused to cooperate, fearful that finding it would fit Sheba's prophecy, so she had him cast into a dry well until he gave up the location.

On the right side in the bottom register is the scene showing Constantine in his tent, dreaming of an angel who told him that if he displayed the Christian cross in battle he would ride to victory. A viewer at the front left side of the choir may have been able to see the ensuing battle carrying on across the adjoining right side wall.

And on the left is the Annunciation. This self-contained image depicts one of the most important events in the Catholic faith, celebrated yearly with a feast on May 25th, and it was a popular subject of art during the Renaissance. Much has been done to work this scene into the logic of the Legend narrative. It is only an issue, however, if Piero's commission was to follow Varagine's narrative as a whole. If not, Piero's problem – the authority of narrative order – dissolves. It makes much more sense in this light, for example, to suggest that this scene is present because Pope Nicholas IV conceded an indulgence in 1290 to the church of San Francesco with respect to prayer in chapel on the day of the Annunciation (see Lavin 1994, 9).

Taken out of holistic expectations, the window wall thus offers less-privileged viewers their own stories. Even a glance of the images of the prophets, the hiding and recovery of the wood and the dream of Constantine draws a remote spiritual world into ordinary experience. While a small, learned audience among the friars could see these in the context of Varagine's Legend, or other versions, other members of the clergy and the laity might associate them with more widely shared oral traditions. Indeed, in terms of engaging an audience, Varagine's spare account almost required it. The plot of the legend of the true cross describes a great story in the tradition of medieval romance novels, with the holy wood as a hero among villains, involved in great quests and battles, as in the stories of the holy grail. No wonder these tales were so popular.

It is probably an archaeologist's bias, but perhaps the thrill of seeing these events in the life of the cross was not so much how they fit the teachings of the church as it was the more practical dream of one day seeing relics of the cross in Florence Cathedral or the Basilica di Santa Croce in Gerusalemme in Rome (see Baert 2004).

The story on the window wall could also address audiences' issues unrelated to either the wonder of the images and their stories, or their harmless teachings, and instead focus on more serious issues. Of the three pictures about the legend on the window wall, the picture of Judas being hoisted up from the well is a more telling inclusion. It follows Varagine's account, but

Piero had many choices as to which part of Helena's quest to depict, and Gaddi omits this scene altogether. Like the battle scenes, it may be here as another depiction of Christianity in struggle.

Life for Jewish communities in Tuscany was always uneasy. Less than 50 years before Piero began his work, it was decreed in Florence on January 24, 1406 that

> Considering that Jews or Hebrews are enemies of the cross, of our Lord Jesus Christ, and of all Christians, and that they engage in usury against the command of Holy Church … no Hebrew or Jew … shall be allowed … to lend money at usury … in the city, *contado*, or district of Florence, or in any city, region or place under the jurisdiction of the Commune of Florence, under penalty of 1,000 florins.
>
> (Brucker 1998, 240–41)

The Florentine tactics were hypocritical, because the wealthy needed the expertise and resources of Jewish financial experts. Jewish people living in Arezzo at this time had important roles as doctors and bankers (e.g., Cassuto 1918, 34–35) and, after the 1406 decree, the Florentines agreed to allow bankers in Arezzo to continue their work with payment of an annual tax (Salvadori and Sacchetti 1990; Debby 2000, 190–91). There were several modifications and other decrees thereafter, including one in 1463, three years before Piero's commission ended, which placed limitations on the size of the Jewish population and required resident Jews to wear a yellow letter O "on the left breast, over the clothing in a visible place" (Brucker 1998, 242).

In the church of San Francesco, this issue may have been as important as the more distant wars, and both more important than fidelity to the legend. Whether or not Piero chose this scene himself, the image of Judas being hauled out of the well was another victory in the battle for Christianity; it demonstrated popular prejudice and religious partisanship unencumbered by the more complicated relations between wealthy Aretine merchants and their Jewish bankers.

Releasement from narrative

From the point of view of audiences within the choir and chancel, the Legend emerged differently as seen from different areas of the church, but none obeyed the rule of narrative, which in the Legend would be a continuous flow between Adam's death and Heraclius' humble bearing of the true cross to Jerusalem. The episodes were part of the larger story to those who knew it – but looking at pictures in the chancel could never be a journey through Varagine's summary text, any more than active life follows any of the schemes and plans intended for it.

Look again at Piero's treatment of Seth's meeting with the archangel and the gift he brought from God. Despite its significance in the story, Piero's cursory treatment is no less distanced than Varagine's. In his text, at this seminal moment in the story, Varagine matter-of-factly provides the reader with three different accounts of what Seth received from the archangel – the oil of mercy, a branch to plant on the Mount of Lebanon [Deuteronomy 3:25] or the fruit or seeds "of the tree that Adam ate of" – without weighing their merits, and then goes on with the story. In this, he acknowledges that history came from individual authorities, rather than consensus, and so while it might be said that there was an eternal truth, it was not one that even historians could grasp.

The pressure to resolve situations with a coherent system seems in comparison like an entirely modern preoccupation. Unlike the viewers who look at the images presented to them in the church and walk away, the canonical views of the church and the ideologies within other institutions may have an inertia that extends over centuries with no need to go back to the source at all.

In an archaeology, however, the need for the direct evidence of artifacts and their relations demands that local matters preside, and so, facing the walls of the chancel, our method is more in keeping with an audience who looks at the art rather than ponders ideas it might provoke. Without ideas about central organization, narrative continuity and all the other governing representations in which Piero's work might reside, what is left are the traces of the painter in his local environment, weighing demands, expectations, existing traditions and personal interests and goals and, depending on the circumstances he found himself in on the scaffold, making secondary adjustments to a popular story.

Standing in the church today, these adjustments still exist in the paintings, and it seems best to speculate from a position close to the artist's situation. Even though the terrazzo screens were stripped out long ago, one can think directly about Piero's sense of audience and the practical dilemmas of managing attention to his displays, as he worked around the visual obstacles and constraints that the lay congregation faced. As the situations of ordinary life are not the same as constructs about them, it is possible that, in practice, the highly structured ideological environment implied in church teachings and in its physical architecture was not as rigid as it might seem. Mindful of the teachings of Gregory and Bonaventure in the defense of images as pictures for those who cannot read, and considering Piero's depiction of the sacred Legend with reference to its audiences, it seems pointless in the 1450s if only the friars in San Francesco could see the whole of his works. There may be many situations where hard boundaries become more permeable, as Cooper (2014, 59) observes:

> It is unrealistic to think that lay patrons would never have enjoyed access to the monuments they commissioned … These sacred interiors could be divided for the purposes of preaching and the liturgy, with screens, doors, and veils providing theatrical framings for ritual. At other times, though, they must have been remarkably open arenas, where both friars and laity could experience a range of artistic riches and where many barriers were permeable.

Painters should also be released from their histories and authorities to exist in their works. Piero as a painter faced a cascade of situations in his commission. He chose, or was instructed, to depict events long told that the Bishop of Genoa, Jacopo de Varagine, had summarized in his Latin text of the *Legenda Aurea*. He strayed from a conventional order, perhaps because he did not set out to paint that way or because the situation changed over the years of his commission. In that case, he would have adjusted his plan accordingly and, most importantly, done so outside the authority of the narrative.

There was probably little gravity in these decisions. For most of the people who attended the church, the walls did not display a pictorial narrative that Varagine called the Invention of the True Cross and the Exaltation of the True Cross, or show evidence of any consciousness of nascent science or humanities or other schemes that had not yet arrived in the presentations of history. Piero was not bound to tell *a whole story* at all. Indeed, with respect to the window wall, many brief experiences conjure up worlds – a quick glance, snatches of conversation, touches or senses or, for an archaeologist, finding artifacts – and it should be the same for those who may see only these three extraordinary scenes of the Legend.

Encounters in any part of daily life reach awareness because the perception is direct and sheer experience educates the mind and senses to the available world. In the moments of active perception of a picture, and imagining the life the images resource, the viewer is briefly free of narrative expectations and perhaps closer to an artist's intent.

In the basilica of San Francesco in Arezzo today, the view is democratic, without the medieval screens and furniture in place, and tourists and scholars crowd into the chancel and look up at Piero's work, restored and lit by electric light. Some are content to take in their astonishing presence, but others will look over the walls as an archaeologist scans the ground, searching the images of people, movements and events that meet, or confound, their own expectations.

This literal grounding leads us into the final step along a crooked path through situations and attitudes about art. We began with a mountain and we will end with what seems to be its direct and equally curious opposite, a landscape that an artist painted at least partly in the open air, at the brink of another abyss.

Note

1 See https://www.wga.hu/html_m/p/piero/2/index.html for high-resolution colour images.

References

Alberti, Leon Battista. [1435] 2013. On Painting. In *On Painting: A New Translation and Critical Edition*. Translated and edited by Rocco Sinisgalli. Cambridge: Cambridge University Press.

Allies, Mary H. 1898. *St. John Damascene. On Holy Images. Followed by Three Sermons on the Assumption*. Translated from the original Greek. London: Thomas Baker.

Baedeker, Karl. 1909. *Central Italy and Rome Handbook for Travellers*. London: T. Fisher Unwin.

Baert, Barbara. 2004. *A Heritage of Holy Wood: the Legend of the True Cross in Text and Image*. Leiden: Brill.

Bailey, Martin. 2021. *Van Gogh's Finale: Auvers and the Artist's Rise to Fame*. London: Frances Lincoln.

Baxandall, Michael. 1972. *Painting and Experience in Fifteenth Century Italy*. Oxford: Oxford University Press.

Brucker, Gene, ed. 1998. *The Society of Renaissance Florence. A Documentary Study*. Toronto: University of Toronto Press.

Brumm, A., A.A. Oktaviana, B. Burhan, B. Hakim, R. Lebe, J. Zhao, P.H. Sulistyarto, M. Ririmasse, S. Adhityatama, I. Sumantri and M. Aubert. 2021. Oldest cave art found in Sulawesi. *Science Advances* 7(3).

Carley, Rachel. 2008. Domestic Afterlives: Rachel Whiteread's Ghost. Special Issue: Interior Atmospheres. *Architectural Design* 78:26–29.

Cassuto, U. 1918. *Gli ebrei a Firenze nell'età del Rinascimento*. Firenze: Tip. Galletti e Cocci.

Cooper, Donal. 2014. Experiencing Dominican and Franciscan churches in Renaissance Italy. In *Experiencing Sanctity*, edited by Trinita Kennedy, 47–62. Frist Center for the Arts. London: Philip Wilson Publishers.

Costall, Alan. 1985. How Meaning Covers the Traces. In *Visual Order, the Nature and Development of Pictorial Representation*, edited by N.H. Freeman and M.V. Cox, 17–30. Cambridge: University of Cambridge Press.

Costall, Alan. 1993. Beyond Linear Perspective: a Cubist Manifesto for Visual Science. *Image and Vision Computing* 11:334–341.

Costall, Alan. 2017. 1966 and All That: James Gibson and Bottom-Down Theory. *Ecological Psychology* 29:221–230.

Da Varagine, Jacopo. [1483] 1900. *The Golden Legend, or Lives of the Saints, as Englished by William Caxton*. Vol. Three. London: J.M. Dent.

Debby, Nirit Ben-Aryeh. 2000. Jews and Judaism in the rhetoric of popular preachers: The Florentine sermons of Giovanni Dominici (1356–1419) and Bernardino da Siena (1380–1444). *Jewish History* 14:175–200.

Duggan, Lawrence G. 2005. Was Art Really the "Book of the Illiterate"? In *Reading Images and Texts. Medieval Images and Texts as Forms of Communication. Papers from the Third Utrecht Symposium on Medieval Literacy, Utrecht, 7–9 December 2000*, edited by Marielle Hageman and Marco Mostert, 63–108. Turnhout: Brepols.

Edgerton, S.Y. 2006. *Brunelleschi's Mirror, Alberti's Window, and Galileo's 'perspectivetube'*. História, Ciências, Saúde – Manguinhos 13 (supplement):151–179.

Field, J.V. 1997. Alberti, the Abacus and Piero della Francesca's Proof of Perspective. *Renaissance Studies* 11:61–88.

Gell, Alfred. 1998. *Art and Agency: An Anthropological Theory*. Oxford: Clarendon Press.

Gibson, Eleanor. 1995. Are We Automata? In *The Self in Infancy: Theory and Research*, edited by P. Rochat, 3–13. New York: Elsevier Science Publishers.

Gibson, James J. 1950. *The Perception of the Visual World*. Boston: Houghton Mifflin Company.

Gibson, James J. 1966. *The Senses Considered as Perceptual Systems*. Boston: Houghton Mifflin.

Gibson, James J. 1978. The Ecological Approach to the Visual Perception of Pictures. *Leonardo* 11:227–235.

Goffman, Erving. 1961. *Asylums: Essays on the Social Situation of Mental Patients and Other Inmates*. Garden City: Anchor Books.

Heft, Harry. 2010. Affordances and the Perception of Landscape: an Inquiry into Environmental Perception and Aesthetics. In *Innovative Approaches to Researching Landscape and Health*, edited by C. Ward Thompson, P. Aspinall and S. Bell, 9–32. London: Taylor & Francis.

Hoffmann, D.L., C.D. Standish, M. García-Diez, P.B. Pettitt, J.A. Milton, J. Zilhão, J.J. Alcolea-González, P. Cantalejo-Duarte, H. Collado, R. de Balbín et al. 2018. U-Th Dating of Carbonate Crusts Reveals Neandertal Origin of Iberian Cave Art. *Science* 2018:912–915.

Holmes, George. 1989. Florence and the Great Schism. *Proceedings of the British Academy* LXXV:291–312.

Ingold, Tim. 2000. *The Perception of the Environment: Essays on Livelihood, Dwelling and Skill*. London: Routledge.

Jeffery, Arthur. 1944. Ghevond's Text of the Correspondence between 'Umar II and Leo III. *The Harvard Theological Review* 37:269–332.

Jung, Jacqueline E. 2006. Seeing through Screens: The Gothic Choir Enclosure as Frame In *Thresholds of the Sacred: Architectural, Art Historical, Liturgical and Theological Perspectives on Religious Screens, East and West*, edited by Sharon Gerstel, 185–213. Washington: Dumbarton Oaks.

King, Catherine E. 1998. *Renaissance Women Patrons. Wives and Widows in Italy c. 1300–1550*. Manchester: Manchester University Press.

Knappet, Carl. 2004. The Affordances of Things: a Post-Gibsonian Perspective on the Relationality of Mind and Matter. In *Rethinking Materiality: the Engagement of Mind with the Material World*, edited by Elizabeth DeMarrais, Chris Gosden and Colin Renfrew, 43–51. Cambridge: McDonald Institute for Archaeological Research.

Lavin, Marilyn Aronberg. 1994. *Piero Della Francesca: San Francesco, Arezzo*. New York: George Braziller Inc.

Leopold, D.A., and G. Rhodes. 2010. A comparative View of Face Perception. *Journal of Comparative Psychology* 124 (3):233–251. https://doi.org/10.1037/a0019460.

McLuhan, Marshall. 1964. *Understanding Media: the Extensions of Man*. New York: McGraw-Hill.

McLuhan, Marshall, and Quentin Fiore. 1996. *The Medium is the Massage. An Inventory of Effects*. Berkeley: Gingko Press.

Molyneaux, Brian L. 1997. Representation and Reality in Private Tombs of the Late Eighteenth Dynasty, Egypt: an Approach to the Study of the Shape of Meaning. In *The Cultural Life of Images: Visual Representation in Archaeology*, edited by Brian L. Molyneaux, 108–129. London: Routledge.

Moro-Abadía, Oscar, and Diego Garate Maidagan. 2010. The Beginnings of European Upper Paleolithic Art: A Critical Review. *North Atlantic Archaeology* 2:1–18.

Mossakowski, Stanisław. 1968. Raphael's "St. Cecilia". An Iconographical Study. *Zeitschrift für Kunstgeschichte* 31:1–26.

Mulan, Father Elder, trans. 1914. *The Spiritual Exercises of St. Ignatius*. New York: P.J. Kenedy.

Muntz, Eugene. 1882. *Raphael. His Life, Works and Times*. London: Chapman and Hall.

Oerter, Herbert L. 1968. Campaldino, 1289. *Speculum* 43:429–450.

Polanyi, Michael. 1970. What Is a Painting? *The American Scholar* 39:655–669.
Salvadori, R.G., and G. Sacchetti. 1990. *Presenze Ebraiche Nell'Aretino dal XIV al XX Secolo*. Florence: Olschki.
Sartre, Jean-Paul. 1948. *The Psychology of Imagination*. Secaucus: The Citadel Press.
Setton, Kenneth. 1978. *The Papacy and the Levant 1204–1571. Volume II: The Fifteenth Century*. Philadelphia: American Philosophical Society.
Seybolt, Robert Francis. 1946. Fifteenth Century Editions of the Legenda Aurea. *Speculum* 21:327–338.
Shklovsky, Viktor. [1917] 1965. Art as Technique. In *Russian Formalist Criticism. Four Essays*. Translated by Lee T.Lemon and Marion J. Reis, 3–24. Lincoln: University of Nebraska Press.
Shklovsky, Viktor. 2007. *Energy of Delusion: A Book on Plot*. Translated by Shushan Avagyan. London: Dalkey Archive Press.
Sontag, S. 1977. *On Photography*. New York: Farrar, Straus & Giroux.
Stubblebine, James. 1969. Giotto and the Arena Chapel Frescoes. In *Giotto: The Arena Chapel Frescoes*, edited by J.H. Stubblebine, 71–100. New York: W.W. Norton & Co.
Tsao, Doris Y., Winrich A. Freiwald, Roger B.H. Tootell and Margaret S. Livingstone. 2006. A Cortical Region Consisting Entirely of Face-Selective Cells. *Science* 311:670–674. doi:10.1126/science.1119983.
Turner, David. 1990. The Politics of Despair: The Plague of 746–747 and Iconoclasm in the Byzantine Empire. *The Annual of the British School at Athens* 85:419–434.
Vasari, Giorgio. [1543] 1912. *Lives of the Most Eminent Painters Sculptors & Architects*, translated by Gaston DuC. Deverb. Volume III. London: Philip Lee Warner, Publisher to the Medici Society.
Venturi, Robert, Denise Scott Brown and Steven Izenour. 1972. *Learning from Las Vegas*. Cambridge: MIT Press.
Welish, Marjorie. 1996. Jasper Johns. *Bomb* 57. https://bombmagazine.org/articles/jasper-johns/.
Whiteread, Rachel. 2014. *Transcript of portion of online video interview*, National Gallery of Art, published May 30, 2014,6:6–6:22. https://www.nga.gov/collection/art-object-page.131285.html#relatedpages.
Zinken, Jorg, and Alan Costall. 2019. Diversity and Invariance in Human Social Action: Karl Duncker's "Situational Meanings" and the Schema of Linguistic Relativism. *Paradigmi* 3/2019:491–506. doi:10.30460/95140.
Zollner, Frank. 1992. Ogni Pittore dipinge sè. Leonardo da Vinci and "automimesis". In *Der Künstler über sich in seinem Werk, Internationales Symposium der Bibliotlieca Hertziana*, edited by Matthias Winner, 137–160. Weinheim: VCH.

8 Landscape, energy, ecology

The coarser grounds of fine art

In a book on situations, there can be no conclusion – only more situations. In the same year that a Mi'kmaw incised images of family members, 1844, the English painter J.M.W. Turner (1775–1851) exhibited a painting at the Royal Academy in London that depicts a steam train crossing a railroad bridge in driving rain.

Both these acts fit within a general synchronicity in the 19th century that involved indigenous people and colonial conquerors, and the workers, managers and elites in the industrial countries of western Europe and North America. For this and other global patterns to make sense, they sweep up people in their representations, enroll them in social groups and give them collective minds. However, they do not represent lived life and the people in it, who provide the raw data through their acts. Importantly, all interpreters, whatever they profess or how they approach this problem, are among these individuals, all facing the metaphorical abyss between direct experience and its descriptions and meanings.

From where we began in a gulch, trying to make sense of a scatter of artifacts, to where we are now, standing in front of a landscape painting in an art museum, we have explored ways of using direct experience in analysis at the scale of individuals, local communities and surroundings. As an approach, it starts down the path to ecologies of art in archaeology.

In an ecology of art, every work of art or display, whether a handprint or a rock painting or a huge fresco on the wall of a church, records a sequence of direct actions. Like a performance, it begins to take shape long before the cascade of situations ends with the act of painting. Visual art has an advantage over other performances because the actions of the artist persist in the material outcome, the analogue image. This means that visual art is its own recording, and the record is necessarily perfect. It therefore presents a field of possibilities for an archaeological analysis.

Of course, the problem with the perfect analogue is that it outlives its maker and loses its original meaning and intent. For example, when Vincent Van Gogh laid down his tools after he painted a self-portrait in 1889, his part

DOI: 10.4324/9781351053105-8

in the work ended, and he understood this abandonment. In a letter to his brother Theo, on or about September 20 that year, he added a note urging him to study it:

> Today I'm sending you my portrait of myself, you must look at it for some time – you'll see, I hope, that my physiognomy has grown much calmer, although the gaze may be vaguer than before, *so it appears to me.* [emphasis added]
>
> (Van Gogh [1889] 2009: Letter 805, lines 123–28)

Van Gogh's request provides a reason to rescue the painter's intent for posterity. When a picture enters a world beyond an artist's control, what prevents it from being a mutable brand, commodified for the ideology of the time, is continued attention to the work on the canvas, the painter's intent. Van Gogh protected his own portrait as best he could with his words and, in them, recognized its fragility as a record of a life. Still, his voice persists as a presence and a correction to more distanced and disinterested appropriations of the image.

A return in general to the study of the lives of actual people as lived is increasingly urgent in a world dominated by ideology. For example, in Van Gogh's case, what seems most important today about the once-living man who painted pictures is something he never intended, to be regarded as an ideological resource – as in these comments in a catalogue in a recent exhibition of his work (Tate Britain 2019):

> Today, we look beyond the idea of Van Gogh as a solitary artist. We appreciate his work in vivid and changing dialogue with the culture and ideas of his times and those that have followed.

Overviews are basic urges in any form of navigation through life, and we could not interpret the particular without them. From this remote position, however, finding patterns that suggest collective forces at work is too easy, and the risk is that artists may become mere vectors for ideological trends and, by this, effectively disappear. At least in art museums, however, the original paintings survive their texts. In many other museums, ideology has taken over, and the few artifacts are there to illustrate ideas and opinions. It is therefore crucial that art and archaeology remain grounded, as the richness of the past ultimately depends on its associations with a lived-in world.

A self-portrait seems to be an ideal subject for a grounded interpretation, but it seems impossible to say the same for a mid-19th century landscape painting – because it is supposed to represent something it does not depict. However, a painter who paints a picture *en plein air* uses the world as its own model and, from what our tentative ecology of art suggests about human–environment interaction, painting an outdoor scene on the spot not only picks up visible details, but also some of the active energy in the surroundings.

As a way of closing, we will briefly discuss this notion in reference to J.M. W. Turner's landscape painting, "Rain, Steam, and Speed – The Great Western Railway", which he showed first at the Royal Academy in London in 1844.[1] While this picture is an imaginative view mounted on a flat wall in an art museum, we will attempt to free it from its ideological home and ponder the situation in which Turner might have painted it, by a bridge over the Thames, west of London.

Rain, steam and speed

At first glance, the picture shows a steam engine pulling open carriages filled with people speeding out of a roiling atmosphere of light and colour, created with vigorous swirls, daubs and smears of pigment, across a bridge over a placid river in driving rain. Only one other feature penetrates through this haze, a few arches of a stone bridge in the background that disappears into the masses of pigment on the far shore of the river.

This is not abstraction, meant only to appeal to the senses. Turner creates a clear path to the place he pictures in the caption, inviting viewers to witness the forces of nature and technology, embodied in rain, speed and steam, at a bridge along the railway. His motivation was clear. Everyone at the time knew that the opening of the Great Western Railway was about to change life in southern England. Small steam railways had been carrying coal and a few passengers in the industrialized north in the late 18th century, but the plan of the Great Western Railway was to connect London with the major seaport of Bristol. It would carry freight and passengers at impossible speeds across the countryside in all seasons, easily overcoming horse-drawn coaches, barges and ships. A trip lurching in a stagecoach over often rough and dubious turnpike roads over days would take hours on the rails (see Rosevear 2004).

Turner knew that some of his audience would study it closely. He placed things to grasp in the miasma of colour and light outside the force field of the

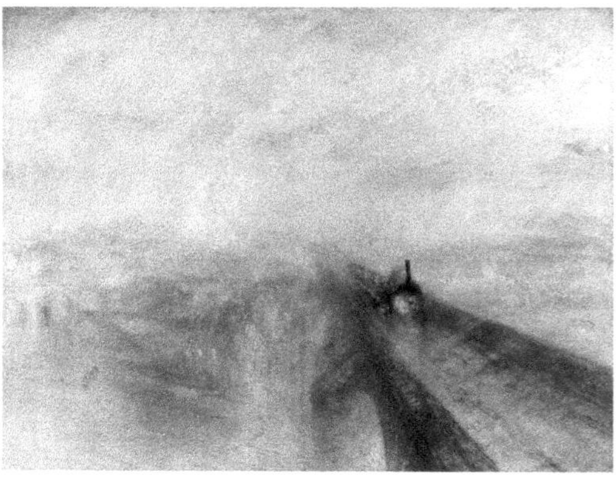

Figure 8.1 Rain, Steam, and Speed – The Great Western Railway

speeding train. Two people drift in a skiff on the river, one hunched under an umbrella; a ploughman guides his horse team forwards through a nearby field despite the downpour; several people on the far shore near the embankment seem to raise their arms in surprise, astonishment or celebration; and, barely visible on the bridge, in the right foreground between the tracks, a hare stretches out in a full gallop ahead of the train.

There is also a ghost here. In the foreground, where the track disappears into the viewing space, a dark and dominant ocherous wash extends away from the viewer along the north side of the bridge and drops downwards at the central pier; once seen, it may resolve into the head and neck of a shadowy horse.

Turner never spoke or wrote about the experience at the bridge, but John Ruskin records an anecdote about Turner getting drenched while looking out of a window in a train from Exeter to Paddington in 1843 (Cook and Wedderburn 1908, 598–601), and some critics still think that he planned the painting from this experience (see Carter 1997). However, if he only took in the scene on the bridge as a passenger, he probably could not see it from the train or, if he did, it was the briefest of glances. William Makepeace Thackeray, reviewing the work in 1899, imagined it speeding at about 50 miles an hour (Thackeray 1899, 439–40), which is about 72 feet per second, and even if the actual train was going half that speed, Turner would have had little time to ponder the scene, wet or not.

Outside what cannot be resolved, there is only the painting, which we will interpret as he offered it to his viewers, as an eyewitness account of a steam engine pulling a train of carriages across a bridge over the Thames on the Great Western Railway in the driving rain. The question for the rest of us, who see only the outcome more than 175 years later, is what we might learn from a painting that implies the act of being there, between the opening of the first railway bridge along the line in 1839 and the 1844 exhibition, that may help constrain and enrich meanings taken from the results.

In this regard, it is crucial here to remember that the world the painting depicts did not begin in 1839 with the construction of the first bridge – it ended. Turner was not caught up in the "railway mania", as it developed soon after (see Murdoch 2003), and the tiny hare, about to be overrun by the train, which has become symbolic of the 19th century industrial assault on nature, could not possibly have carried this burden in 1839, as it was only getting started. When his privileged audience first saw the painting in 1844, they probably made a connection to hare coursing, a blood sport in which a hare is released and chased down by horses and dogs, and they might have wondered how the speed of the train compared to greyhounds in the chase (cf. Thomas 2016). Yet, Turner's point by this would indeed become prophetic. Railroads enabled middle class people to travel quickly and cheaply, and hare coursing, which was an aristocratic pursuit, became a hugely popular spectacle in the whole of the country through the 19th century (Russell 2018).

In all, looking at the subject of the picture and its depiction literally, the subtle visual counterpoints of senses and scenes in the picture suggest that since the first bridge late in the 13th century, traditional rural life at this crossing was about to end, and for Turner it did, in a shock that was physical as well as intellectual, the irruption of the natural forces of rain, speed and steam beside a railroad bridge across the Thames sometime between July 1, 1839 and 1844.

To discover what we can of this situation, we begin with the location pictured. The consensus is that the railway bridge is the one that the civil engineer Isambard Kingdom Brunel (1806–1859) built over the Thames at Maidenhead in 1839. The first stretch of the railway line, a distance of about 22 miles from Paddington Station in London to the village of Taplow, opposite Maidenhead on the left bank of the river, had opened in 1838, awaiting the construction of the bridge. The clues to the location and Turner's physical presence there are not on this span, however – they are in the road bridge that he brings out of the haze of distance. This bridge is stone: two large arches over the river have voussoirs, and there seems to be a hint of a balustrade in the wash above. Plainer approach arches follow, and the whole melds into a mass of colour that resolves into a large hill. From these heights, the elevation drops steadily, and across the tracks at the railway's vanishing point the valley floor stretches out in fields, including the nearest one, which the ploughman and horses turn.

This bridge grounds the picture. It closely resembles the bridge crossing the Thames along the Great Western Road (later called the Bath Road and now the A4), the main route from London to Bath, Bristol and the west. It is a Georgian structure of 13 arches, built of Portland stone and brick, which replaced a wooden one in 1777. Cooke and Owen (1811) describe the view from this bridge in a book about the Thames, and it matches Turner's topographic shading in the painting: "To the north is seen the bold range of woody heights, crowned with Taplow, Cliefden, and Hedsor; while to the south the river flows through one unvaried, uninteresting level ..." (1811, 223). And the painting also shows the slight convergence between the two structures, which cross about a half mile east. On this evidence, the picture presents a specific event: a journey by train from London on the Great Western Railway, heading west, on or after July 1, 1839, captured just as it crosses the approach of Brunel's Maidenhead Bridge.

By locating the picture here, Turner defines the capacities and potential of its surroundings, and, in effect, invites viewers to take part in the sensory experience that he did – not as a path through the painting to some higher purpose, but to share the shock of being there, beside the track. To capture the drama of the advancing train, Turner created an illusion to deal with two problems in reckoning from the ground. First, if he actually painted the train as if it was bearing down on him as he stood beside the track, the engine would obscure its string of carriages, and so he floats a few hundred feet off the ground and to the left of the track. Then there is the crucial association of the bridges. The road bridge is actually a thousand feet away, a little too far to convey its mass and detail or balance the composition, so he brings it closer – and preserves its angular strike to the railway bridge. Thus, as one hovers magically above the track and looks east towards the advancing train, a sidelong glance takes in the last two river arches of the road bridge and the approach arches that join the far shore.

Turner's care with the road bridge is more than establishing its place and its role in composition and reinforcing his prophecy. It also emphasizes the intrusion of the railroad over the Thames valley. He was familiar with this crossing long before the railway began, as it was the main route to the west country. In the years 1811 and 1813–15, he traveled there to make sketches for a proposed book and would have crossed over the bridge in a stagecoach along the Bath road. In one of his travels, he probably paused on the bridge and looked south down the Thames over the fields where he may have seen the view that Cooke and Owen (1811) describe, "the summer buildings on Monkey Island, and the tower of Bray church" (1811, 223). All this changed when Brunel built his great brick masterpiece and blocked the view. The painting thus embodies a mixture of regret and anticipation, rather than a single point of view. Brunel's son says in a biography of his father that the aesthetics of the bridge compensated quickly for the "dreaded invasion of the railway" (Brunel 1870, 174) and, indeed, the two great structures remain the focus of attention in the area today.

This invitation to tangible experience, through the reportage of the painting, makes Turner's great work doubly intrusive. As a work of art, it blasts the bland pictorialism of conventional landscapes to announce the irruptions of technological change that would have profound and lasting effects on the entire population. And as an historical record, it provides the slightest touches of local life, bound more clearly to the languid current of the great river in the outskirts of the villages of Maidenhead and Taplow than across the two great roads. As Turner uses these local references to ground the work's intellectual conceits, they may be personal references to the experience of witnessing the train's passage, rather than the effects he knew would come to English society. Perhaps the rain, the leading word in Turner's title, was not simply a painterly device. July had just begun when the first train crossed the Maidenhead bridge in 1839, and it became the seventh wettest July on record in England (Wigley et al. 1984, 14) in a year that would have nearly double the average precipitation for southern England. Perhaps the speeding hare was indeed an aside to some in his audience who traveled along the Bath road to hare coursing events at Ashdown Park in the North Wessex Downs, about 48 miles to the west. And perhaps the pareidolia of the horse, with the train bearing down on it, was actually intentional, rather than a delusion of this writer, and Turner's shadowy prophecy of things to come.

As a painting is a result of a series of acts, it is a convergence of physical and social flows active during its creation, and when the painting is of a place that endures, standing near where the artist stood may provide insights into the work. On the surface, Turner's surroundings are gone. Maidenhead and Taplow have spread to the river. A dense network of roads and buildings have squared up the land and blocked views across it; and the field where the ploughman toiled is in the landscaped grounds of several houses. And just as there is less of Turner's view, there is far too much knowledge about the worlds before and after that singular crossing of the bridge. The Great Western Railway is now an acronym, and GWR trains are faster and coolly efficient, with no smoke, fire or steam, and they cross many times a day; the A4 is noisy with traffic; and the river is busy with pleasure and commercial craft.

Yet, in all the noise and bustle of modern life, some of the energy that the painter picked up still exists in the deep surroundings. When Turner watched from the road bridge, from a train carriage and standing by the tracks, he did so in the thrall of the most fundamental of intersectional forces – the crossing. Art, philosophy and social science are not needed here. Across the world and through time, from the small camps of hunters and gatherers to great cities, settlements grow where diverse flows of energy meet, through the increased travel and resources they afford.

In the 9th century there was a small wharf and ferry across the river here, and the Saxon hamlet nearby took its name – Maiden-Hythe or New Wharf (Nash 2020); in 1280 the first bridge improved travel along the Saltway between London and the primary source of salt, the brine-pits in Droitwich, about a hundred miles northwest of the bridge; and in 1777, the grand Georgian bridge was completed. Nash relates that at the peak of traffic in the 18th century, 90 coaches a day passed over it. Maidenhead thus seemed to have invited Turner's choice, as the flows of water, history and society all converged there, and moments after the train crossed the bridge, the quick, staccato surges of its steam power created fresh metaphors about the yielding of an old, spent world to something apparently new.

In the end, looking back from the Thames to where we began, the meeting of forces at this crossing is repeated in various ways around all the landmarks, rock paintings and carvings and frescoes we have discussed and in large and small measures everywhere. We all pick our sources of attention out of the world, but the work is also in the world and among its forces.

This book has presented some alternative ways of exploring the past that use the world as its own model, especially theories of action and perception in psychology, biology and art, and it has done so with critical attention to the language of description and the nature of the imagination. Philosophy and literature are important in this regard, and two significant works have contributed to this approach to representations outside those already discussed.

In 1485, Thomas Aquinas (c. 1225–1274), in his great compendium of Catholic theology, *Summa Theologica* (2007), pondered whether an angel might occupy a place or whether two angels could occupy the same place. Later scholars mocked the very idea of this as a problem, but it is an entirely practical question about the substance of belief. He concluded that because it is incorporeal, "an angel is said to be in a place which is corporeal, not as the thing contained, but as somehow containing it" (2007, 268). This defines the problem of governing representations, which we have tried to work around in this study. The resistance to such ideological reinforcements comes from an illustrated poem by William Blake (1757–1827), the *Marriage of Heaven and Hell* (composed in 1790). In one vision, he writes that an angel brought him to the edge of the abyss of the five senses (the abyss in our title), where he witnessed huge, terrible spiders and other devils, but then, when the angel left his side, the appearance dissolved, and he found himself on a pleasant bank beside a river.

Working in the field, surrounded by traces of lost people and alive with present communities, archaeologists struggle with the received wisdom of similar authorities, but being close to the earth, they also have a clear idea about what will happen in the history to come. Turner's painting will stay for a time on its wall, and scholars and critics will write about it until attention shifts to something else and, for a much longer time, the bridges will cross the banks of the Thames at Maidenhead, until no-one remembers these names. Even then, as people occupy and fight over new representations, streams and their surroundings will simply draw and sustain cultural life and hold some of its traces for a curious few, and, outside instruction, we will all flow on!

Note

1 See https://www.nationalgallery.org.uk/paintings/joseph-mallord-william-turner-rain-steam-and-speed-the-great-western-railway for a high-resolution colour image.

References

Aquinas, Thomas. 2007. *Summa Theologica, Volume I-Part* I, translated by Fathers of the English Dominican Province. New York: Cosimo.

Brunel, Isambard B.C.L. 1870. *The Life of Isambard Kingdom Brunel, Civil Engineer.* London: Longmans, Green and Co.

Carter, Ian. 1997. Rain, Steam and What? *Oxford Art Journal* 20:3–12.

Cook, E.T., and Alexander Wedderburn, eds. 1908. *The Works of John Ruskin. Volume XXXV, Præterita and Dilecta*. London: George Allen.

Cooke, William Bernard, and Samuel Owen. 1811. *The Thames; or, Graphic Illustrations. Seats, Villas, Public Buildings, and Picturesque Scenery, on the Banks of that Noble River.* London: Printed for Vernor, Hood, and Sharpe.

Murdoch, Jill. 2003. *From Elephant to Penge West: The Railway in the Artist's Landscape.* PhD Thesis, Institute of Railway Studies, University of York.

Nash, David Ford. 2020. *East Berkshire Town and Village Histories.* Wokington: Nashford Publishing.

Rosevear, Alan. 2004. *A booklet on the Turnpike Roads around Reading.* http://www.turnpikes.org.uk/Reading%20turnpike%20roads.htm

Russell, Edmund. 2018. *Greyhound Nation. A Coevolutionary History of England, 1200–1900*. Cambridge: Cambridge University Press.

Tate Britain. 2019. *The EY Exhibition Van Gogh and Britain.* Exhibition catalogue. https://www.tate.org.uk/whats-on/tate-britain/exhibition/ey-exhibition-van-gogh-and-britain/van-gogh-and-britain-exhibition-guide.

Thackeray, William Makepeace. 1899. *May Gambols; or, Titmarsh on the Picture Galleries. Ballads and Miscellanies.* Volume 13 in the "Biographical Edition" of Works, 419–445. London: Smith Elder.

Thomas, Inigo. 2016. The Chase. *London Review of Books* 38, No 20.

Van Gogh, Vincent. [1889] 2009. Letter 805. In *Vincent Van Gogh, The Letters*, edited and translated by Leo Jansen, Hans Luijten and Nienke Bakker. Amsterdam: Van Gogh Museum and Huygens ING. http://vangoghletters.org/vg/letters/let805/letter.html#.

Wigley, T.M.L., J.M. Lough and P.D. Jones. 1984. Spatial Patterns of Precipitation in England and Wales and a Revised, Homogeneous England and Wales Precipitation Series. *Journal of Climatology* 4:1–25.

Index

For Product Safety Concerns and Information please contact our EU
representative GPSR@taylorandfrancis.com
Taylor & Francis Verlag GmbH, Kaufingerstraße 24, 80331 München, Germany

www.ingramcontent.com/pod-product-compliance
Ingram Content Group UK Ltd.
Pitfield, Milton Keynes, MK11 3LW, UK
UKHW020140200926
475702UK00005B/835

* 9 7 8 1 1 3 8 4 8 4 0 7 8 *